旧厂房的空间蜕变
Uptodate in outdated factory
Spatial from Temporal

建筑立场系列丛书 No.17

中文版

韩国C3出版公社 | 编

时跃 陈帅甫 杨宇芳 薄寒光 赵姗姗 葛永宏 胡筱狄 | 译

大连理工大学出版社

资讯

004 无限塔——探求未知_GDS Architects
008 Aliah酒店_Hiperstudio+Arkiz
012 魏玛包豪斯新博物馆_Menomenopiu Architects
016 蝉建筑_Casagrande Laboratory

建筑脉动
速度、雄心和挑战

021 恩佐·法拉利博物馆_Future Systems+Shiro Studio
　　速度、雄心和挑战_Silvio Carta

旧厂房的空间蜕变

043 不同时代的混合体_Silvio Carta
048 建于原马德里屠宰厂里的红牛音乐学院_Langarita-Navarro Arquitectos
060 IMd 工程顾问公司_Ector Hoogstad Architecten
068 红牛公司新总部_Sid Lee Architecture
082 Tesa 105改建项目_Andrés Holguín Torres+David R. Morales Hernández
092 Zap' Ados_Bang Architectes
100 米诺——意大利米利亚里诺青年旅社_Antonio Ravalli Architetti
106 La Fabrique——创新机器_TETRARC

学校重建与复杂性

学校扩建

125 对朴素的学校建筑进行增建_Jorge Alberto Mejía Hernández
130 D.Manuel I中学_BFJ Arquitectos
146 Vila Viçosa高级中学_Cândido Chuva Gomes Arquitectos
156 Francisco de Arruda学校_José Neves
164 拉滕贝格中学_Architekt Daniel Fügenschuh ZT GmbH
172 贝尔纳多特学校的扩建_Tegnestuen Vandkunsten
180 裴斯塔洛齐学校_SOMAA+Gabriele Dongus

188 建筑师索引

建筑立场系列丛书 No.17

News
004 Tower Infinity - Not Seeing is Believing_GDS Architects
008 Aliah Hotel_Hiperstudio+Arkiz
012 New Bauhaus Museum in Weimar_Menomenopiu Architects
016 Cicada_Casagrande Laboratory

Archipulse
021 Enzo Ferrari Museum_Future Systems+Shiro Studio
   Speed, Ambition and Challenge_Silvio Carta

Spatial from Temporal
Uptodate in outdated factory
043 *Temporal Hybrids_Silvio Carta*
048 Red Bull Music Academy in the Old Madrid Slaughterhouse_Langarita-Navarro Arquitectos
060 IMd Consulting Engineers_Ector Hoogstad Architecten
068 Red Bull New Headquarters_Sid Lee Architecture
082 Conversion Tesa 105_Andrés Holguín Torres+David R. Morales Hernández
092 Zap' Ados_Bang Architectes
100 MiNO, Migliarino Youth Hostel_Antonio Ravalli Architetti
106 La Fabrique, Creative Machine_TETRARC

Palimpsest and Complexity
School Extension
125 *Adding on the Modest Monument_Jorge Alberto Mejía Hernández*
130 D.Manuel I Secondary School_BFJ Arquitectos
146 Vila Viçosa High School_Cândido Chuva Gomes Arquitectos
156 Francisco de Arruda School_José Neves
164 Rattenberg Secondary School_Architekt Daniel Fügenschuh ZT GmbH
172 Bernadotte School Extension_Tegnestuen Vandkunsten
180 Pestalozzi School_SOMAA+Gabriele Dongus

188 Index

无限塔——探求未知

韩国仁川Cheongna地区的无限塔作为东北亚全球动态经济的New Gateway地标性建筑,代表了韩国的新"精神",因为它拥护全球社区,而不是只注重自身发展。其设计灵感基于韩国及其人民的自我反省,他们曾经的状况以及最重要的:他们引以为自豪的领域和在全球社区里是如何被大家认识的。从这个意义上讲,竞争可以演变为契机,提供社会评论并质疑所有有限的现状。

在当代,"地标性塔楼"这个词往往表现了一个国家向其他国家展示其经济繁荣和科技成果水平的愿望/意识,以引起其他国家的羡慕。建筑师的内心、思想和动机没有陷入试图与世界上伟大的标志性建筑竞争或发展另一种形式的"最高瞭望塔"的泥沼中,而是要表现不存在的力量和"虚无"的力量,而这些象征意义经常被路易斯·康提及。从这方面来看,无限塔是在其不存在的情况和实际的体量中进行强化的,这是一种自相矛盾的情形。在这个空间中,它以信心和谦卑做着不懈斗争,我们因而发现了人文的曙光。人们只能希望,这种空间最终形成一种情感/精神之旅,超越很快会遗忘的典型现世经历。

随着战后韩国迅速工业化,并且现在被指定为世界第十二大经济体,韩国将凭借其对历史的有力传承和对未来的愿望继续奋斗。虽然高塔引发的兴奋感已经消失了很多,但其所展示出的韩国科技似乎是惊人和不寻常的,从对无限性的本质和对未知可能性的理解角度入手,人们更强调将重点放在真实信息和对人类的贡献中。现在有一个向世界展示的机会,展示出韩国及其人民理想的最重要的方面是对未知的、不可见的事物的不断探求。

建筑师真诚地希望这一项目将象征性地代表韩国人民的"集体灵魂",为下一代提供一个有意义的挑战,并给他们遗留宝贵财富,因为每一个人都在追求有意义的生活,而没有被他们的所想和所见蒙蔽。

Tower Infinity – Not Seeing is Believing _ GDS Architects

Positioned as the New Gateway landmark in the dynamic global economy of Northeast Asia, Tower Infinity in the Cheongna area of Incheon, Korea represents the new "Soul" of Korea by celebrating the global community rather than focusing inward on itself. The design inspiration was based on an introspection of the Korean nation, its people, where they have been and most significantly where they see themselves headed as a proud nation and how they are perceived in the realm of this global community. In this sense the competition evolved as an opportunity to provide a social commentary and challenge the status quo of all which is finite. Too often in modern times, the phrase

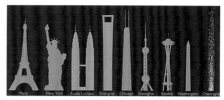
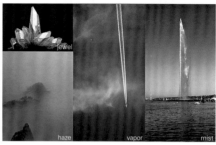

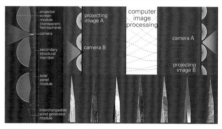

"Landmark Tower" is indicative of a nation's desire/ego to showcase a level of economic prosperity and technological achievement for others to envy. Rather than fall victim with trying to compete with the great iconographic landmarks of the world or develop another version of the "tallest observation tower", architects' heart, minds and motivation were set to represent the Power of Absence and the Strength of "Nothingness" often referred to by Louis Kahn. In this respect, Tower Infinity is paradoxically strengthened in its absence and its volume, and in this void we find the presence of hope for humanity as it struggles with confidence and humility. One can only hope that the result is an emotional/spiritual journey that transcends the typical temporal experience soon forgotten.

With the rapid industrialization of post-war Korea and current designation as the 12th largest economy in the world, Korea continues to struggle with its strong connection to the past and its aspirations for its future. While too many excitement of a tower disappearing showcases Korean technology may seem striking and unusual, a greater emphasis should be placed on the true message and contribution to humanity through the lens of understanding the nature of the infinite and the possibility of the unknown. Here lies an opportunity to showcase to the world that the most important aspect of their ideals as a nation and people is that constant search for the unknown, the unseen.

It is architects' sincere hope that this project will symbolically represent the "collective soul" of the Korean people while providing a meaningful challenge and legacy to the next generation as each individual searches for a life rich in meaning and is not blinded by what they think and see.

项目名称：Tower Infinity
地点：Cheongna, South Korea
设计建筑师：GDS Architects, GDS Korea
主建筑师：Samoo Architects
助理建筑师：A&U Architects
结构工程师：King-Le Chang & Associates
立面/可持续性：ARUP-Hong Kong
甲方：Land and Housing Corporation of Korea
用途：Observation Tower and Cultural village
用地面积：110,425m²
建筑面积：Observation Tower_9,800m², Podium-Cultural Village_49,989m²
景观面积：6,832m²
总楼面面积：10,325m²
建筑覆盖率：25.4%　楼面比率：54.1%
最大高度：450m
结构：Steel Diagrid
外部饰面：BIPV, Glass, Steel
停车场：1,804 cars
预计竣工时间：2014

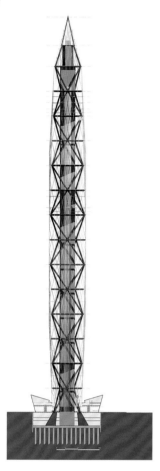
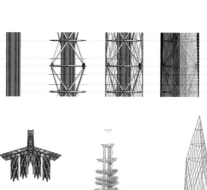

brace core, link truss

1st + outer ring truss

2nd + diagrid frame

斜肋架构框架
diagrid frame

内部框架(核心)
inner frame(core)

外部框架(管)
outer frame(tube)

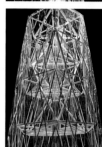

+405m roof
+397m infinite sky-lounge
+387m sky-high observation
+381m global bistro
+371m sky-group observation
+366m refuge/mechanical room

+232m soaring theater
+222m pronto cafe
+216m middle observation/experience Cheongna
+211m refuge/mechanical room

roller coaster platform
U-CITY control room

road | wedding park | lake | adventure park | center park | aqua park | canal

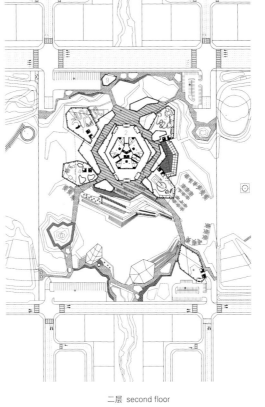

level 397

level 381

level 364

level 232

level 216

level 209

level 17

level 12

一层 first floor

1 水晶水上公园	1. crystal water park
2 水晶塔	2. crystal tower
3 冒险区域	3. adventure land
4 水族馆	4. aquarium
5 雨林/蝴蝶体验区	5. rainforest/butterfly experience
6 探索博物馆	6. discovery museum
7 水晶剧场	7. crystal theater
8 婚礼大厅	8. wedding hall

二层 second floor

1 水晶水上公园	8 水晶湖桥	3. adventure land
2 塔	9 信息中心	4. aquarium
3 冒险区域		5. discovery museum
4 水族馆	1. crystal water park	6. event island
5 探索博物馆	2. tower	7. island walk
6 项目岛		8. crystal lake bridge
7 岛上人行道		9. information center

Aliah酒店

该项目赢得了Aliah（一家推广可持续发展的公司）举办的竞赛。

本次竞赛的目的是为2014年世界杯开发一个可持续性豪华酒店综合建筑项目，这个酒店将要建在圣保罗郊外。

1. 该酒店作为人与环境和谐相处的空间

城市化和工业化进程促进了社会的形成，为现代人创建了大量令人惊叹的设施。另一方面，这也牺牲了人类与自然和环境的诸多关系。随着这一进程的发展，这种分离关系通过污染的加剧、自然资源不可避免的缺乏和大城市中心的生活质量下降而被人们感知。

对环境可持续性问题越来越多的认识为促进人与自然环境之间和谐发展的重要举措提供了机会。

2. 风景作为主角

Aliah项目的目的不仅是建造一座舒适的创新型酒店，还要建造一个超自然的空间，以鼓励游客对更美好、更健康、更平衡的生活方式进行重新审视。Aliah项目提议的建筑旨在将这些概念作为重点，恢复那些作为人类和环境之间相互作用的基础的必要而永恒的价值。

沿着抵达轴线排列的线性广场瞭望台是主要的设计元素，因为它对酒店和会议中心的入口进行了协调，并总结出与周围景观结合的建筑经验。

3. 转化为可持续性的建筑语言

建筑师选择将酒店综合建筑设置在沿场地脊线的纵轴的较高地块附近，这样使其本身既能适应自然地形，又能欣赏到美丽的景观。本地区的敏感性分析反映在建筑物与地面的相互作用上，形成了一种辩证关系，在这种关系中，建筑为沉思冥想提供了物质支持。

服务于酒店房间的开放式走廊围绕一个郁郁葱葱的中心庭院，突出了植被，这样，它就不只具有单纯的连接功能，还成为树冠间的观光区。

该项目的建筑语言旨在突出所采用的可持续性和环境舒适性的不同策略，展示了包括获得高环境性能在内的解决方案，同时确保能源效率和资源节约。

Aliah Hotel _ Hiperstudio + Arkiz

This project has been awarded as the winner of a competition organized by Aliah, a company which promotes sustainable development.

The goal of the competition was to develop a project of a sustainable luxury hotel complex for the 2014 World Cup, to be built in the outskirts of Sao Paulo.

1. The hotel as a space of rapprochement between man and environment

The processes of urbanization and industrialization that led to the formation of our society produced an amazing universe of facilities for the modern man. On the other hand, it sacrificed much of his relationship with nature and the environment. As

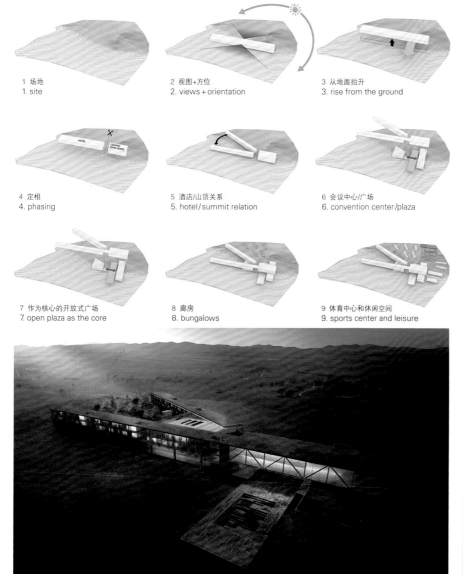

1. 场地
1. site

2. 视图+方位
2. views + orientation

3. 从地面抬升
3. rise from the ground

4. 定相
4. phasing

5. 酒店/山顶关系
5. hotel / summit relation

6. 会议中心/广场
6. convention center / plaza

7. 作为核心的开放式广场
7. open plaza as the core

8. 廊房
8. bungalows

9. 体育中心和休闲空间
9. sports center and leisure

this process intensifies, the effects of this detachment are felt through increased pollution, the inevitable scarcity of natural resources and loss of quality of life in large urban centers.

The growing awareness of the issue of environmental sustainability provides opportunities for important initiatives to promote rapprochement between man and natural environment.

2. The landscape as the protagonist
The Aliah project aims to be not only a comfortable and innovative hotel but

项目名称：Aliah Hotel
地点：São Paulo, Brazil
建筑师：Hiperstudio, Arkiz
项目团队：Alexandre Hepner, Joao Paulo Payar, Matheus Marques, Rafael Brych, Ricardo Felipe Goncalves
用地面积：75,000m²
建筑面积：12,230m²
总楼面面积：7,620m²
设计时间：2012
竣工时间：2014 (planned)

1. convention center + exhibition
2. access plaza
3. hotel
4. grove
5. bungalow
6. sports and leisure center
7. parking lot

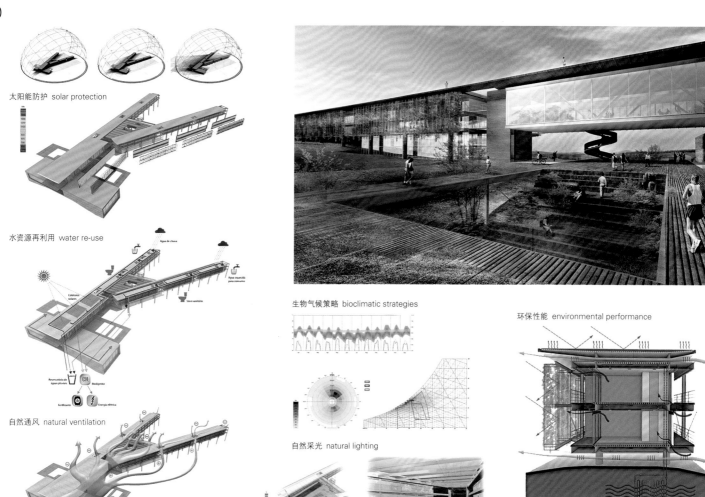

太阳能防护 solar protection

水资源再利用 water re-use

自然通风 natural ventilation

生物气候策略 bioclimatic strategies

自然采光 natural lighting

环保性能 environmental performance

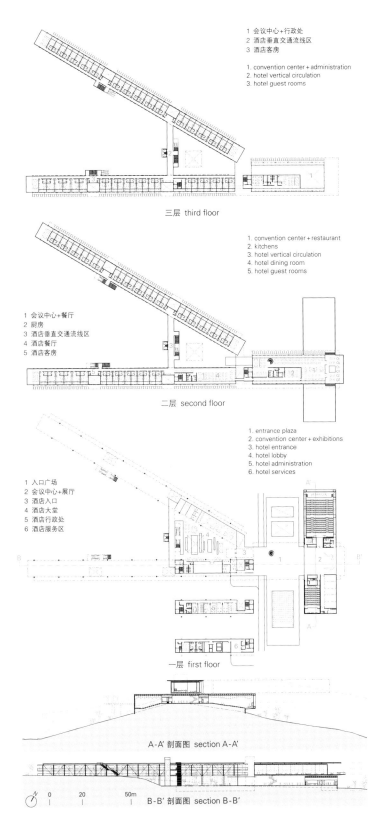

1 会议中心+行政处
2 酒店垂直交通流线区
3 酒店客房

1. convention center + administration
2. hotel vertical circulation
3. hotel guest rooms

三层 third floor

1. convention center + restaurant
2. kitchens
3. hotel vertical circulation
4. hotel dining room
5. hotel guest rooms

1 会议中心+餐厅
2 厨房
3 酒店垂直交通流线区
4 酒店餐厅
5 酒店客房

二层 second floor

1. entrance plaza
2. convention center + exhibitions
3. hotel entrance
4. hotel lobby
5. hotel administration
6. hotel services

1 入口广场
2 会议中心+展厅
3 酒店入口
4 酒店大堂
5 酒店行政处
6 酒店服务区

一层 first floor

A-A' 剖面图 section A-A'

B-B' 剖面图 section B-B'

also a transcendental space where visitors are encouraged to re-evaluate their attitude in favor of a better, healthier and more balanced way of life. The proposed architecture for the Aliah project seeks to put these concepts into focus, restoring essential and timeless values that underlie the interaction between humans and the environment.

A linear plaza-belvedere, arranged along the arrival axis, is a strong element as it coordinates the access to the hotel and convention center and leads to the discovery of the architectural experience integrated with the surrounding landscape.

3. Architectural language of translating sustainability

Architects chose to place the hotel complex near the higher ground in a longitudinal axis along the land's ridge, accommodating itself on the natural topography and providing beautiful views towards the landscape. The sensitive analysis of this region is reflected in the way in which the building interacts with the ground, establishing a dialectical relationship in which architecture acts as a physical support for contemplation.

The open corridors that serve the rooms of the hotel embrace a central leafy patio, emphasizing the presence of vegetation, transcending the mere function of connection to become a tour through the treetops.

The architectural language of the project seeks to highlight the different strategies of sustainability and environmental comfort liabilities that were employed, demonstrating the solutions involved to achieve high environmental performance while ensuring the energy efficiency and resource savings.

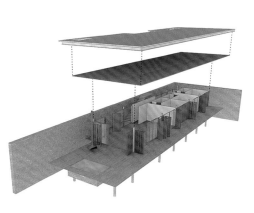

魏玛包豪斯新博物馆

2012年3月16日，魏玛古典基金会宣布了为魏玛包豪斯新博物馆项目举办的国际建筑竞赛的结果。

评审团在世界范围内为魏玛包豪斯新博物馆项目开展的建筑设计竞赛评选出了两名二等奖和两名三等奖获得者。陪审团还评选出了三名优秀奖获得者。获奖者名单的公布也使建筑设计竞赛正式画上了句号，来自世界各地的536家建筑事务所参加了此次竞赛。三名优秀奖获得者之一是总部位于罗马/巴黎的Menomenopiu建筑事务所。

评审团的报告显示："透明度使内外部之间具有连续性，内外浑然一体。空间的折衷主义风格设计非但没有使建筑物成为一种障碍，反而使其成为城市中一种强有力的连接体，与一条跨越包豪斯悠久历史的灵活的线性道路相交。创造一个新整体的构思通过多种艺术和运动的结合来实现。"

这座建筑被构想为一个露天广场，位于新旧城市和公园这三种主要城市力量的交汇点，是一座可以在内部和四周进行不同活动的灵活"建筑"。

覆顶的广场作为一个多功能的空间，根据开放时间来决定是否与上层展览楼层进行互动，多功能空间与公园形成自然联系，并对城市开放。

作为建筑关键因素的"智能屋顶"十分灵活，能够利用现代技术来满足未来建筑物的需求。屋顶图案产生的光影变化催生了Josef Albers的绘画作品。

50%的屋顶（1250m²）覆盖了光伏板，每天能提供250kW的能量，满足博物馆60%的能源

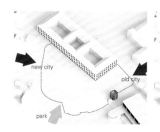  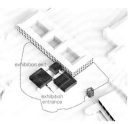 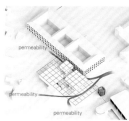 

消耗需求。滑动百叶窗调控着室内不同季节的光照条件,满足不同箱形空间的功能性需求。

"箱中箱"系统与技术型屋顶一起,保证在不同季节都能对气候进行最好的控制。

"我对多层透明表面上重叠图像的理念很感兴趣。"

——出自Moholy-Nagy的《明亮空间的调节器》

New Bauhaus Museum in Weimar
_ Menomenopiu Architects

On March 16, 2012, the Klassik Stiftung Weimar(Weimar Classics Foundation) announced the results of the international architectural competition for the New Museum in Weimar.

The jury awarded two second-place and two third-place prizes in the world, opening architectural design competition for the New Bauhaus Museum in Weimar. The jury also conferred three Honorable Mentions. The announcement of the winners officially concludes the architectural design competition, in which 536

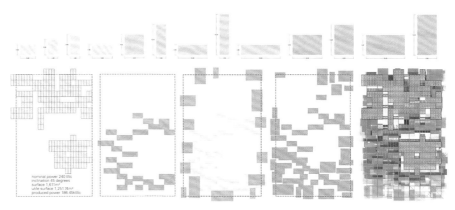

architectural offices around the world participated. One of the three Honorable Mentions was the entry by Rome/Paris-based Menomenopiu Architects.

From the Jury Report: *"The transparency makes the continuity between inside and outside, the outside comes inside and vice-versa. An eclectic conquest of the space makes the building not a barrier but a dynamic urban link, crossed by a linear flexible path across Bauhaus history. The idea of creating a new unity is achieved by the marriage of many arts and movements"*.

The building is conceived as an open square at the crossing point of the three main city forces, old and new city and the park, a flexible "object" that allows different activities inside and around it.

A covered square as a multifunctional space that may or may not interact with the upper exhibition floor, depending on opening hours, a multitasking space opened to the city in a natural relation with the park.

The "clever cover", as a key element is flexible to accommodate contemporary technologies to fullfill the future needs of the building. The game of light and shadows made by the pattern of the cover evokes the drawings of Josef Albers.

The 50% of the roof (1250m²) is covered by photovoltaic panels which provide up to 250kW/day, satisfying 60% of the museum energy consumption. Sliding shades manage the seasonal solar interior contributions, satisfying programmatic needs of the different boxes.

The system "box in the box", in synergy with the technological roof, guarantees the best climatic control through the different seasons.

"I'm interested in the idea of overlapping images over several layers of transparent surfaces."
– Moholy-Nagy's *Light Space Modulator*

项目名称：New Bauhaus Museum in Weimar
地点：Weimar, Germany
建筑师：Menomenopiu Architects
设计团队：Alessandro Balducci-Gilberto, Bonelli-Mario Emanuele, Salini-Rocco Valantines
合作者：Francois Zab-Marco, Lavit Nicora-Marco, Conti Sikic-David, Yahn-Drahi-Luca Stortoni
工程：Marc Hammon- Giulia Fatarella
景观：Bassinet Turquin Paysage 图像：+imgs
用地面积：10,618m² 建筑面积：3,200m²
总楼面面积：4,295m²

A　　B　　C　　D　　E　　F

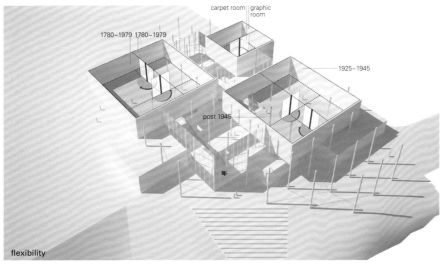

flexibility

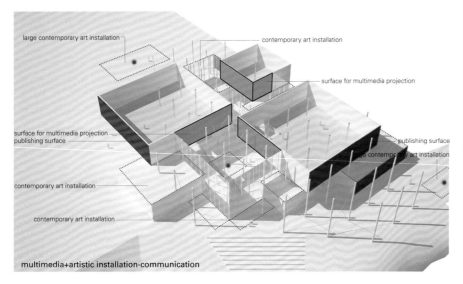

multimedia+artistic installation-communication

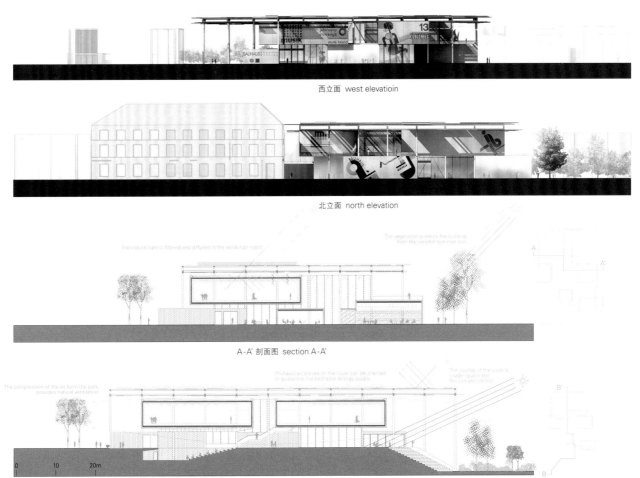

西立面 west elevatioin

北立面 north elevation

A-A' 剖面图 section A-A'

B-B' 剖面图 section B-B'

蝉建筑

蝉建筑是现代台北市机械结构中的一种有机空间,是工业"昆虫"进行后工业变形的茧状空间。该建筑风格基于当地人性尺度的灵活竹结构知识,包含了一种高水平的即兴创作和"昆虫"理念——开放形式。

蝉建筑坐落在台北市中心一处待开发的场地内。同时,它充当了周围地区的公共空间,并作为大学工作室和其他自发活动的休息区——公共空间。

进入蝉建筑内部,人们就感觉到周围的城市消失了。茧是室内空间,但完全对外部空间开放——它在呼吸、振动,既柔软又安全。这个空间将"吞噬"现代人,为其提供穿越千年之旅的可能性,使其意识到事物都是一样的。蝉建筑是一种"昆虫"建筑,其空间是一种公共空间。

蝉建筑是为台北市进行的一次城市针灸,穿透工业建筑懒散的坚硬表皮,以触到原始地面,并与集体智慧(即连接台北盆地中人与自然的当地知识)接触。蝉建筑的茧状空间是现代人与现实之间偶发的调节物。没有比自然更现实的东西。

Cicada _ Casagrande Laboratory

Cicada is an organic void in the mechanical texture of modern Taipei, a cocoon for post-industrial metamorphosis of industrial insects. The architecture is based on the local knowledge of human scale flexible bamboo structures containing a high level of improvisation and insect mind – open form.

项目名称:Cicada
地点:Taipei City, Taiwan
建筑师:Marco Casagrande
项目团队:Frank Chen, Yu-Chen Chiu, Shreya Nagrath, Arijit Sen
项目管理:Delphine, Peng Hsiao-Ting / JUT Group , Nikita Wu / C-LAB
尺寸:34m long, 12m wide, 8m high
室内空间:270m²
材料:bamboo, broken concrete, broken glass, steel, soil, creepers
竣工时间:2011
摄影师:©AdDa (courtesy of the architect)

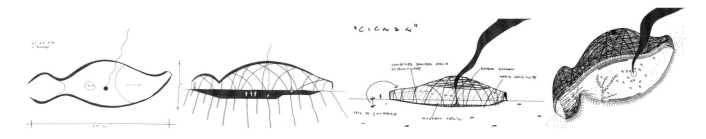

The Cicada is situated on a site in central Taipei waiting for development. Meanwhile it acts as a public sphere for the surrounding neighborhood and as a lounge for university workshops and other spontaneous activities – public space.
As one enters the Cicada, the surrounding city disappears. The cocoon is an interior space but totally opens outside – it is breathing, vibrating, soft and safe. The space will swallow the modern man and will offer him a possibility to travel a thousand years back in order to realize, that things are the same. Cicada is an insect architecture and the space is a public sphere.
Cicada is urban acupuncture for Taipei city penetrating the hard surfaces of industrial laziness in order to reach the original ground and get in touch with the collective Chi, the local knowledge that binds the people of Taipei basin with nature. The cocoon of Cicada is an accidental mediator between the modern man and reality. There is no other reality than nature.

速度、雄心和挑战
Speed, Ambition

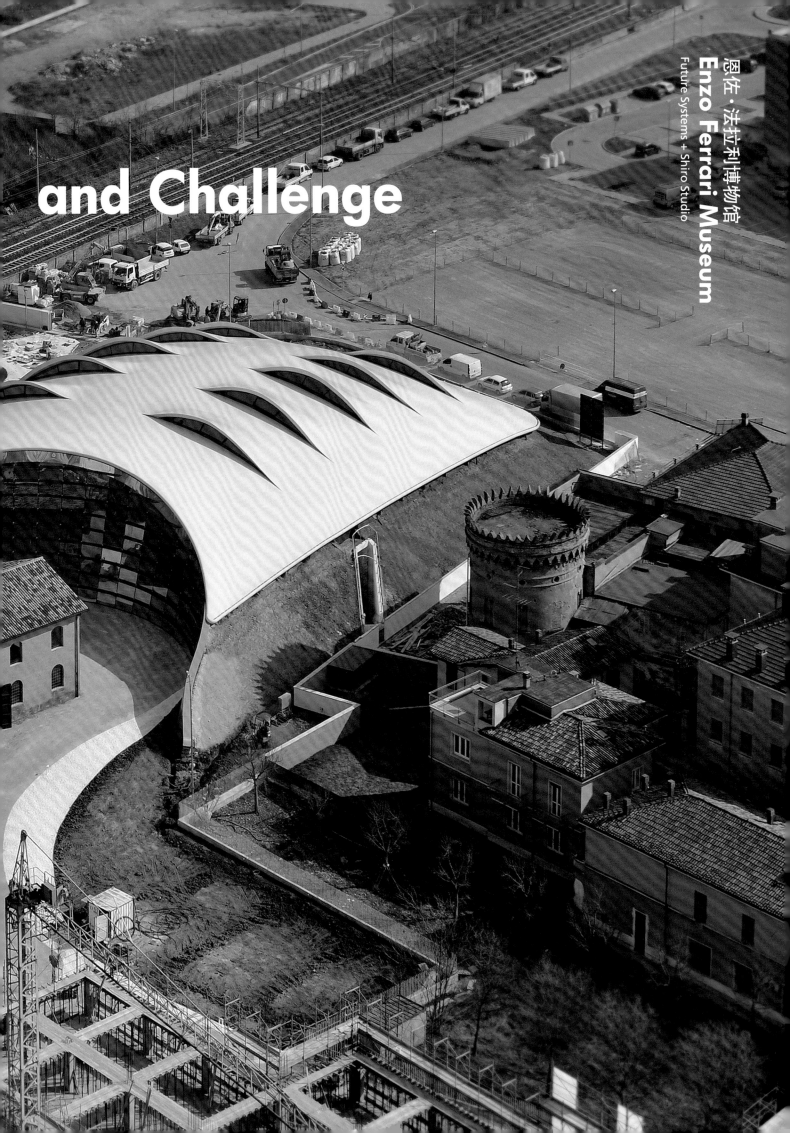

and Challenge

恩佐·法拉利博物馆
Enzo Ferrari Museum
Future Systems + Shiro Studio

速度、雄心和挑战
Speed, Ambition and Challenge

这个博物馆是将新与旧、过去与未来融合的具体化建筑。除了恩佐·法拉利这个更为重要的人物,我更容易联想到简·卡普利茨基,他的过去以及我所在的四郎工作室,我们共同完成了多年梦想的未来。

——Andrea Morgante, 2012年

恩佐·法拉利

法拉利这个名字使人们立即联想到世界上最豪华的汽车品牌。在意大利,尤其是在公司创始人恩佐·法拉利的故乡,这个名字也与一个人的故事有关,这个人的智慧和对赛车的热情使其获得了巨大的成功。事实上,恩佐·法拉利的一生成为所有有抱负的企业家的成功典范,其故事寓意当一个人一辈子朝着一个目标努力工作后就会取得世界范围内的成功。法拉利开始在一家小型汽车公司工作,随后参加各种赛车比赛,直到1929年他成立了法拉利斯库德里亚赛车队。1988年,90岁高龄的法拉利去世。在他的带领下,法拉利斯库德里亚赛车队从一个小品牌转变成为世界知名品牌,并且在国际赛事中赢得无数奖项。除了获得几个荣誉学位和奖项,法拉利死后被列入国际赛车名人堂(包括那些对赛车做出了重大贡献的人,他们可以是车手、制造商、开发商或者工程师)。

纳塔尔住宅

对法拉利创始人的传记所产生的兴趣引起了人们对以下方面——汽车以及与法拉利本人有某种联系的场地的重大关注,例如,法拉利的诞生地——纳塔尔住宅——一座19世纪初建造的住宅,它与附近的车间一起,是恩佐·法拉利出生和成长的地方,也是他的故事的发源地,在这里,他迈出了作为汽车制造商和赛车手的第一步。纳塔尔住宅被恩佐·法拉利出售,以购买他的第一辆赛车,这座住宅时常被遗弃,部分年久失修,直到2003年,恩佐·法拉利基金会决定将其融入新博物馆项目中。该项目是区域层面中更广的理念的一部分,目的在于提升所规划的艾米利亚-罗马涅区(即该项目中所称的意大利Motar山谷),这一地区是玛莎拉蒂、兰博基尼或者杜卡迪品牌的发源地。

古老的住宅和新博物馆

将纳塔尔住宅融入新项目的想法在一开始就成为将来介入建筑要

This Museum is the materialization of old and new, past and future. Beside the more prominent figure of Enzo Ferrari I like to think of Jan Kaplický, his past, and myself (Shiro Studio) completing the future that we dreamt for so many years. – Andrea Morgante, 2012

Enzo Ferrari

The name Ferrari is immediately associated with the most luxury car brand throughout the world. In Italy, and especially in the native area of the company's founder, Enzo Ferrari, the name is also associated with the story of a man whose intelligence and passion for race cars have led to enormous success. The story of the life of Enzo Ferrari is – in fact – a significant example for all aspiring entrepreneurs, serving as a parable of the worldwide success which can eventually come after a lifetime of effort and work toward a single objective. Ferrari began by working in a small car company, and subsequently raced for various stables, until he founded the Scuderia Ferrari racing team in 1929. He passed on at the age of 90 in 1988 after having transformed his small beginnings at Scuderia into a world-renowned brand, winning numerous prizes in international races. Besides several honorary degrees and awards, Ferrari was posthumously appointed to the International Motorsports Hall of Fame, which encompasses those who have contributed significantly to the racing world with their accomplishments, either as race drivers, manufacturers, developers or engineers.

The Casa Natale

The interest in the biography of Ferrari's founder has brought significant attention to those objects, cars and places somehow related to his persona. The Casa Natale, Ferrari's birthplace, for instance – an early nineteenth century house, along with its adjoining workshop – represents the origin of Enzo Ferrari's story as the place where he was born and raised, and took his first steps as a car manufacturer and driver. The Casa was sold by Enzo Ferrari in order to buy his first race car, and it proceeded to be occasionally abandoned and in a state of partial disrepair until 2003, when the Enzo Ferrari Foundation decided to include it in its new museum project. That project is part of a broader idea at the regional level which is intended to promote the area in which several brands, such as Maserati, Lamborghini or Ducati, were founded, under a proposed definition of Emilia-Romagna as the Motor Valley of Italy, as it has been called within the project.

The Old Casa and The New Museum

The idea of integrating the Casa Natale into the new project has thus represented a fixed point for future interventions since the very beginning. Moreover, it should be observed that the northeast periphery of Modena, where the Casa sits, is characterized by quite typical 3–4-storey compact buildings with pitched roofs clad in terracotta tile, conferring the built environment a reddish-brownish tone. Moreover, *"the existing built environment has*

坚持的理念。此外，人们观察到，纳塔尔住宅所在的摩德纳东北边缘地区，以比较典型的3~4层的带有陶土瓦斜屋顶的紧凑建筑为特色，它们赋予建筑环境红棕色的基调。此外，"原有的建筑环境在过去的三年里已经发生了巨大的变化，其平均质量因现代的、平庸的、无志向的建筑而面临着挑战。当你开发这片区域时，这种挑战是非常显而易见的"，Future Systems建筑事务所的原总监Andrea Morgante解释说。

观察这张原始图片，Future Systems建筑事务所针对博物馆竞标的优胜方案可能会显得相当大胆，甚至与环境不和谐。然而，对这座大楼的解读还需要更深入的观察。我们应该回归到对恩佐·法拉利的性格方面，观察他的远见、胆识和勇气等方面。一位像他这样的赛车手，不论是在他与其他汽车制造商（福特或阿尔法·罗密欧）的工作关系中，还是在与其他赛车手的比赛中，都必须不断应对一些非常规的挑战，并联系他所处的环境做出坚定的决定。胆识、勇气和远见的结合可能是他获得广泛成功的秘诀。新博物馆一致地反映出这些价值。这种结合在Morgante概念中十分明确："新大楼……的精神、语言和材料与要展示的车产生共鸣。
"事实上，"他继续说，"对眼前的环境有一个深刻的认识，即将法拉利

changed dramatically in the past three years, with the average quality being challenged by modern, mediocre architecture, with no ambition. This is quite visible as soon as you explore the area", explains Andrea Morgante, former Director of Future Systems.
Looking at this original picture, the winning proposal of Future Systems for the Museum Competition can appear quite daring, and even as being in dissonance with the context. However, the reading of this building requires deeper observation. We should return to the character of Enzo Ferrari and see his visionary, bold and daredevil aspects. A racing driver such as him must have been continuously tackling challenges, unconventional and firmly determined with regard to his context, both in his work relationships with other car manufacturers (Ford or Alfa Romeo) and in his races against other drivers. This combination of boldness, bravery and vision may have been the recipe for his wide-ranging success. The new museum quite consistently reflects such values. This association is quite clear in Morgante's vision: *"the new building [...] resonates in spirit, language and materials with the cars which is intended to showcase. "As a matter of fact,"* he continues, *"there is a profound awareness of the immediate context, assuming the old Ferrari House as the centre of gravity for the whole operation. We wanted to create a sensitive dialogue between the two exhibition buildings that showed consideration for Ferrari's early home and underscored the importance of the museum as a unified complex made up of several elements."*

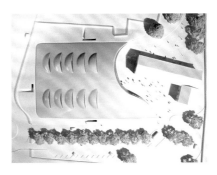

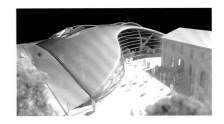

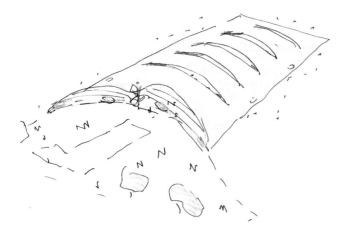 

老住宅作为整个操作过程的重心。我们希望在两座展览建筑间建立起一种敏感的对话,这两座展览建筑要考虑到法拉利早期居住的住宅,同时强调博物馆作为一座由几个要素构成的统一建筑综合体的重要性。"

从这个角度来看,新博物馆很明显以更深刻的方式来反映其环境:通过将恩佐·法拉利变为品牌核心的几个原始因素具体化。因此,新大楼外表光滑、线条清晰,在环境中易于识别。这种将生活方式转为建筑外形的尝试体现出简·卡普利茨基设计的优点和重要性,而这种体现从早期草图(详见其第一张草图,设计于2004年的竞标中)就已开始。

然而,这座博物馆如同一辆设计得非常好的跑车,其质量不仅表现在它的外观和气动外形上。其漂亮的外壳背后隐藏着复杂而又极其巧妙的技术。屋顶就是这一特征的典范。事实上,它的大小和形状本身就是一个挑战,就像Morgante解释的那样:"以前从没有建造过3300m²的双曲面铝质屋顶,设计尤其要考虑到的是屋顶体系基本上是由6m长的挤压模块形成的。由于这项技术在此前从未如此大规模应用,因此便产生了一些我们在设计阶段无法预见到的技术挑战。除去这种复杂性,最终的屋顶忠实于原来的设计,即没有明显接缝的连续性立体屋顶。"

此外,"视觉语言形成了立体的黄色外壳,"Morgante解释说,"这种语言是对汽车设计的致敬和颂扬。参观者甚至在进入大楼前,就能够本能地意识到这座博物馆的性质和使命。"

此外,立面一方面在新大楼的整体外观上发挥着重要作用,同时随着建筑对地热能源的利用,立面在整个项目的气候控制方面也做出了重要贡献。事实上,鉴于新大楼部分位于地下,Morgante解释说,"地热系统使我们无需使用任何空调就能够设计和建造博物馆(美术馆和法拉利住宅展馆),考虑到夏季摩德纳高达34°C的严峻气候,这是个了不起的成就"。所有设计,加上光伏技术和水循环系统的使用,使建筑成为一台复杂的机器。

博物馆的另一个基本方面,是外部营造的氛围和内部设计的氛围之间的对比。外部——基于所蕴含的形式语言,不需要说教式的阐释("没有必要在外面放置一块大牌子"),内部为游客提供一种完全不同的体验:"内部环境更加中立、飘渺。内部景观与展出17辆漂亮汽车和

Read through this lens, it becomes quite plain that the new museum is thus reflecting its context in a deeper way: by giving shape to those original aspects that turned Enzo Ferrari into the core of the brand. The new building is hence slick, sharp and definitely recognizable in its context. In this effort of translating a life approach into building shapes there lie both the merit and the importance of Jan Kaplický's design, beginning with the early sketches (see his first sketch, created in 2004 during the competition).

However, like an extremely well-designed sport car, the quality of the museum does not end with its appearance and aerodynamic shapes. Behind its beautiful shell, the museum hides complex yet extremely smart technology. The roof is exemplary of this inclination. Its size and shape, in fact, have represented a challenge in themselves, as Morgante explains: *"Building a 3,300 m² double curved aluminum roof was something that was not done before, especially keeping in consideration that the roof system is essentially formed by a modular of 6 meters long extrusion. As this technology has never been applied before on such a large scale, this created several technical challenges that we could not foresee during the design stage. Despite this complexity the final result was faithful to the original vision of a continuous, three-dimensional roof with no visible joints."*

Moreover, *"the visual language used to form the tridimensional yellow shell,"* explains Morgante, *"is a tribute and a celebration to car-design. The visitor, even before entering the building, is instinctively aware of the nature and vocation of this museum."*

Also, the facades play an important role in the overall appearance of the new building on one side, while contributing significantly to the climate control aspects of the entire project, along with the building's use of geothermal energy. Indeed, given that the new building is partially underground, *"the geothermal systems,"* explains Morgante, *"allowed us to design and build both museums (the Gallery and the Ferrari House exhibition) without using any air-conditioning, which is quite an achievement considering the severe climate in Modena, where in summer the temperature can reach 34 degrees."* All this, in addition to the employment of photovoltaic technology and water recycling systems, render the building a complex machine.

Another fundamental aspect of the museum is the contrast between the atmosphere created on the outside and that designed for the inside. While the outside – based on the formal language implied – requires no didactic explanation (*"no need to place a big sign on the outside"*), the inside offers visitors a completely different experience: *"Once the inside environment becomes more neutral, ethereal. The interior landscape is now responding to the necessity*

一些文件的必要性相呼应。内部的连续性和流动性默默地支持着展出的汽车和其他物体。在白色的空间中,一切都似乎漂浮起来,光从玻璃立面和屋顶天窗进入,使整个空间充满大量的自然光",Morgante解释说。

相互关系

除了新博物馆建筑和原有城市内容(这种原有的城市内容就我们所见,是通过更进一步地思考法拉利创始人的性格来进行阐述的)之间的关系,建筑的增建部分和纳塔尔住宅之间的关系仍然值得深思。19世纪30年代由恩佐·法拉利的父亲建造的法拉利住宅与周围的城市环境相契合,与之相比,博物馆增建结构显然具有不同比例、主要形状、颜色和材料。

新建筑与原有建筑存在三个方面的联系:摩德纳黄色铝质屋顶的高度与纳塔尔住宅坡屋顶的高度相同;呈弯曲状的新展览空间平缓地围绕在房子周围;而且其规模和轮廓与原有的住宅相比,似乎占绝对优势。对于这种选择,Morgante清楚地解释道:"新展览大楼的特色之处是为住宅和车间提供了框架。"我们可以由此观察到上述关系,即Morgante定义为"和平与尊重"的关系,已经被建筑师以鲜明而独具匠心的方式提出来,而这座古建筑也借此将博物馆的过去以及法拉利神话的起源紧密联系在一起,同时新大楼承担起了在未来拓展法拉利品牌的责任。

Morgante解释说:"新旧大楼的关系也许是整个增建过程中最重要的方面,恩佐·法拉利在1930年出售自己的住宅,购买他的第一辆赛车,这是他一生中最特别的一段插曲。这种纯粹的由勇气和决心驱使的行为使他开始了他的职业生涯。新大楼必须展现出对未来的决心的姿态。因此,新画廊是最现代的技术的实体化展现,就像法拉利赛车一样。

在物质关系方面,我们希望以某种方式将新旧大楼联系在一起,向这座象征性建筑致敬。"

恩佐·法拉利似乎是恰好立于博物馆两部分之间,这两个部分分别代表了同一个长故事的起源和未来。Future Systems建筑事务所似乎已全面完成这项艰巨的任务,在形状和空间上再现了恩佐·法拉利的性格,并在事务所的政策面前也显然优先考虑了这些因素。

of displaying 17 beautiful cars and several documents. The continuity and fluidity of the interior are silently supporting the cars and the objects on display. Everything seems to be floating in a white vacuum, filled with generous natural daylight, entering the building from the glass facade and the roof sky-lights", explains Morgante.

Mutual Relationships

Besides the relationship between the new museum building and the existing urban content, which – as has been seen – is resolved by means of a deeper consideration of the character of Ferrari's founder, the architectural connection between the new section and the Casa Natale still merits contemplation. Whereas the Ferrari House – built in the 1830s by Enzo's father – fits quite comfortably in the urban surroundings, the new part of the museum is manifestly different in terms of proportions, main shape, color and materials.

The new building relates to the existing at three points: the height of the Modena yellow aluminum roof is the same as that of the pitched roof of the Casa; the sinuous shape of the new exhibition space gently curves around the house; and moreover its size and footprint shape – as compared to the existing Casa – seem overwhelming. This choice is clearly explained by Morgante: "*The new exhibition building dramatically frames the house and workshop.*" We can thus observe that the above-mentioned relationship, defined by Morgante as "*peaceful and respectful,*" has been set forth by the architects in a clear and distinctive way, whereby the old building anchors the museum to the past and the origins of Ferrari's myth, while the new building appears to take on the responsibility of projecting the ambitions of the brand in the future.

"*The relationship between the old and new,*" explains Morgante, "*is perhaps the most crucial aspect of the whole intervention. Enzo Ferrari sold the house in 1930 to buy his first racing car, and this is an extraordinary episode in his life. A sheer act of bravery and commitment started his career. The new architecture had to reflect this gesture of commitment towards the future. The new Gallery therefore is the materialization of the most contemporary technology available, just like his cars.*

In terms of physical relationship we wanted the new building to somehow gently relate to the old house, paying respect to such a symbolic building."

The persona of Enzo Ferrari seems to be standing directly between the two parts of the museum, which ideally represent the origin and the future of the same long story. Future Systems seems to have fully accomplished this difficult task, representing in shapes and spaces the character of the man, apparently putting these considerations before company policy. *Silvio Carta*

2004年，Future Systems建筑事务所赢得了为意大利摩德纳设计一座新博物馆的国际竞赛。博物馆致力于展示恩佐·法拉利（1898—1988年）的赛车传奇和从商经历，包括一座位于19世纪初的住宅内的展览空间（这位赛车巨头的出生地和成长地），还包括附近的车间以及一座新建的独立展览大楼。

简·卡普利茨基在2009年去世之后，Future Systems建筑事务所办公室被解散。Andrea Morgante是Future Systems建筑事务所的原总监，现在是四郎办公室的总监，被委任监督博物馆的建造。全面修复的住宅和车间提供了由Morgante设计的额外展览空间。

黄色的雕刻铝质屋顶带有十个切口，外形刻意与汽车引擎盖上的进气口类似，能够进行自然通风和采光，同时表现出汽车设计的审美价值。屋顶为双曲铝质结构，面积为3300m^2，是第一次如此大规模地应用这种铝结构。建筑师与造船者（他们对有机雕刻形式和防水功能的熟悉使其成为理想的合作伙伴）和建筑覆层专家合作，最终建筑外形使用榫槽和凹槽系统安装的铝板组合构建而成。屋顶明亮的摩德纳黄色是法拉利公司的色彩，也成为法拉利徽章（一匹奔腾的马）的背景颜色。黄色也是摩德纳的官方色彩。

卡普利茨基想在两座展览大楼之间建立一种敏感的对话，展示出设计考虑了法拉利的早期家庭因素，并强调博物馆作为几个要素组成的统一复杂结构的重要性。面向新展览大楼的视野形成了住宅和车间的框架，同时，从住宅和车间的外面望去，人们就立刻能体会到这座新展览大楼的功能和内部设施。新展览大楼的高度最高可达12m——与住宅高度相同——其体量延伸到地下。此外，新大楼平缓地围绕在住宅周围。

玻璃立面在平面图中是弯曲的，倾斜12.5°。每个窗格由预张拉的钢索支撑，且能承受40t的压力。这些窗格和缆索的技术规范意味着设计在获得最大功能的同时，还使立面具有更大的透明度。在夏季，热传感器刺激立面上的窗户和屋顶，使冷空气流通。主展览大楼内部的一半体量位于地下，地热能源用于为建筑供暖和制冷。它是意大利第一座使用地热能源的博物馆。大楼还采用了光伏技术和水循环系统。

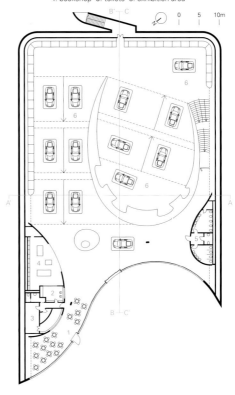

一层——画廊 first floor_gallery

1 自助餐厅/餐厅　2 售票柜台　3 厨房
4 书店　5 卫生间　6 展览区
1. cafeteria/restaurant 2. ticket desk 3. kitchen
4. bookshop 5. toilets 6. exhibition area

地下一层——画廊 first floor below ground_gallery

1 会议室　2 教学室　3 录像装置室
4 员工室　5 技术室　6 储藏室　7 机械设备间
1. conference room 2. didactic room 3. vedio installation room
4. staff room 5. technical room 6. storage 7. plant room

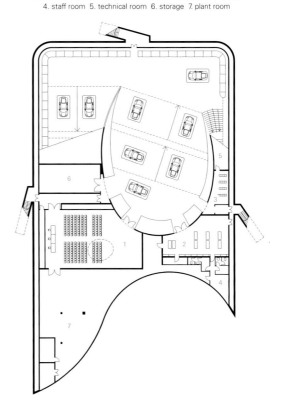

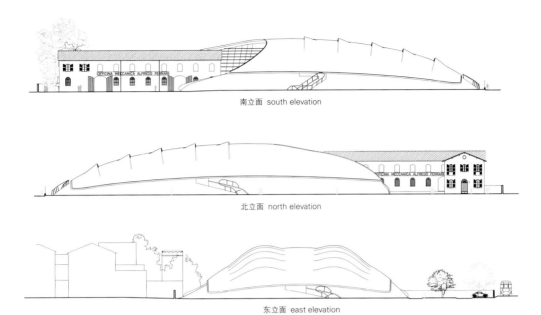

南立面 south elevation

北立面 north elevation

东立面 east elevation

进入新大楼的游客可以将整个展览空间一览无遗：这是一个大型的、开放的白色房间，墙壁和地板间的过渡十分微妙，它们可被作为一个单一的表面。一个伸展的半透明膜将光线均匀地分散开来，薄膜与狭缝（从一侧延伸到另一侧，允许空气逸出，并产生了罗纹效果）相结合，使人想起汽车内部的设计语言。书店和咖啡馆位于入口的一侧，而其他设施位于入口的另一侧。它们的屋顶都涂了摩德纳黄色，并且采用了水泡状吊舱的形式。坡度平缓的斜坡逐渐将大楼周围的游客从地面引到地下一层，Morgante设计的展示台设置在交通路径上。这些展示台将汽车托举到高于地面45cm的地方，因此人们可以从不同的角度观看，并作为艺术作品来欣赏，而不是简单地放置在房间内的物体。这个开放式空间一次最多可展示21辆汽车。备用的展览材料陈列在沿墙壁放置的皮箱中。坡道的底部和入口的正下方是视听室，成为永久性展览的一部分。视听室旁边是灵活的教学空间和会议室，它们都带有雕刻的洞口，使人们可以看到入口区域。

In 2004 Future Systems won an international competition to design a new museum in Modena, Italy. Dedicated to motor racing legend and entrepreneur Enzo Ferrari (1898–1988), the museum comprises exhibition spaces within the early nineteenth century house where the motor racing giant was born and raised, and its adjoining workshop, as well as a separate, newly constructed exhibition building.

Following the death of Jan Kaplický in 2009, the office of Future Systems was dissolved. Andrea Morgante, formerly of Future Systems and now director of Shiro Studio, was appointed to oversee the museum's completion. The fully restored house and workshop provide additional exhibition space designed by Morgante. The sculpted yellow aluminum roof with its ten incisions – intentionally analogous to those air intake vents on the bonnet of a car – allows for natural ventilation and day lighting, and expresses

项目名称：Enzo Ferrari Museum
地点：Via Paolo Ferrari 85, Modena, Italy
建筑师：Jan Kaplický (Future Systems)
项目建筑师：Andrea Morgante
参赛团队：Jan Kaplický, Andrea Morgante, Liz Middleton, Federico Celoni
项目团队：Andrea Morgante, Søren Aagaard, Oriana Cremella, Chris Geneste, Cristina Greco, Clancy Meyers, Liz Middleton, Itai Palti, Maria Persichella, Filippo Previtali, Daria Trovato
艺术指导：Andrea Morgante (Shiro Studio)
画廊展览设计：Jan Kaplický (Future Systems), Andrea Morgante (Shiro Studio)
恩佐·法拉利住宅展览空间设计：Andrea Morgante (Shiro Studio)
项目管理：Politecnica-Modena
结构、机械、电气设计，环境影响评估，健康和安全负责人：Politecnica
技术总监：Giuseppe Coppi (CdC – Modena)
主要承包商：Società Consortile Enzo
甲方：Fondazione Casa Natale Enzo Ferrari
用地面积：10,600m²
概念设计时间：2004 竣工时间：2012
摄影师：
©Andrea Morgante (courtesy of the architect)
-p.28^{bottom}, p.30, p.32, p.33, p.34^{bottom}, p.36~37, p.40~41
©David Pasek (courtesy of the architect) -p.37
©Studio Cento29 (courtesy of the architect)
-p.20~21 p.25, p.28^{top}, 21, 24^{top}, p.35, p.38, p.39

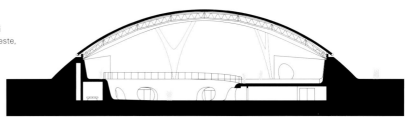
A-A' 剖面图 section A-A'

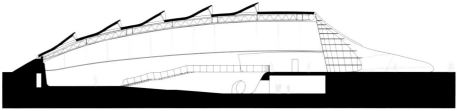
B-B' 剖面图 section B-B'

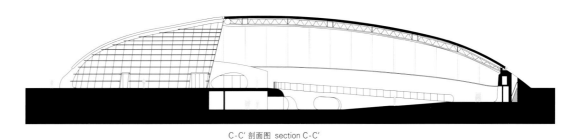
C-C' 剖面图 section C-C'

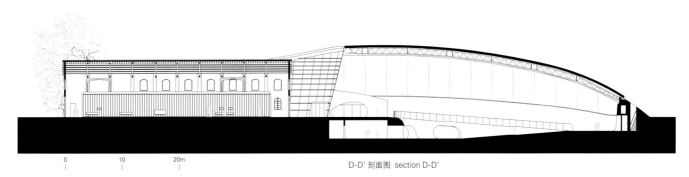
D-D' 剖面图 section D-D'

1. estrusioni fre-curvate in alluminio, spessore 2.5mm, larghezza 120mm, verniciata secondo specifiche tecniche 2. profilato in alluminio a sezione circolare, d=30mm, passo ogni 120 3. graffa di connessione in alluminio, 30x30cm 4. doppio strato di isolamento in vetro cellulare 8+8cm, classe ignifuga 0, resistente al vapore con cuaina bituminosa continua impermeabilizzante 5. lamiera di tamponamento in alluminio verniciata in nero opaco RAL 9004 6. infisso in vetrocamera apribile verso l'esterno manualment. vedi prospetto 7. lamiera grecata in acciaio zincato. microfrata, con lana di roccia nelle grecature. spessore TOT. 137mm 8. profilato in alluminio a sezione circolare, d=30mm 9. graffa di connessione in alluminio, 30x30cm 10. centina verticale di supporto in mmetallo, spess. xxcm 11. bordo calandrato doppoamente curvato in alluminio, spees. 2mm, verniciato secondo le specifiche tecniche della copertura 12. tamponamento in terno in lastra di alluminio verniciata secondo specifighe copertura, spessore 2mm 13. connessione centine con fazzoletto metallico e bulloni. riferirsi alle tavole dello strutturista per spessori e diametri 14. profilo metallico di bordo lucernaio 15. vuoto sull'interno 16. montante superiore infisso vetrato. 17. doppio strato di isolamento in vetro cellulare 8+8cm, classe ignifuga 0, resistente al vapore con guaina bituminosa continua impermeabilizzante. questo strato poggia sulla lamiera grecata portante. 18. estursioni in alluminio, spessore 2.5mm, larghezza 120mm, verniciata secondo specifiche techiche 19. centina secondaria 20. costola trasversale in metallo. spess. xx 21. profilo metallico UPN 320-riferirisi alle tavole strutturali

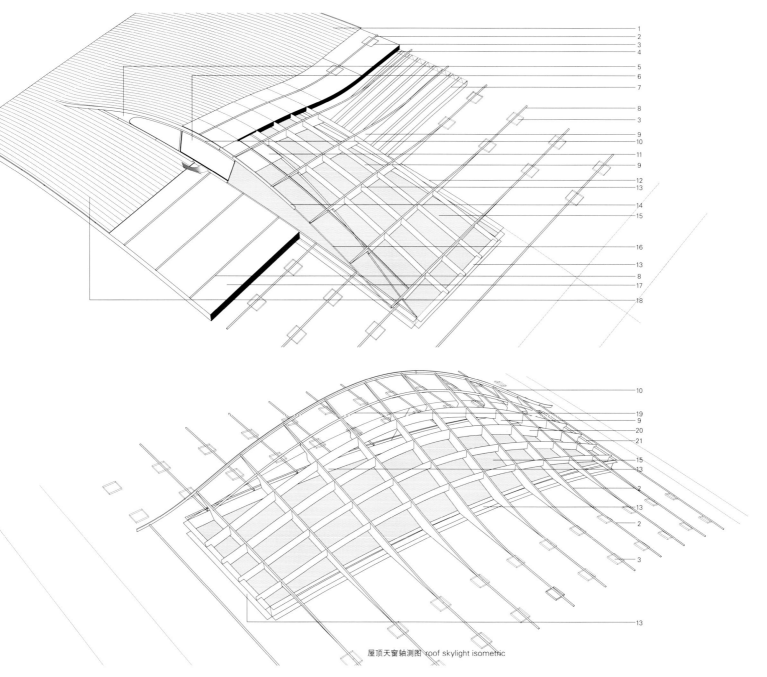

屋顶天窗轴测图 roof skylight isometric

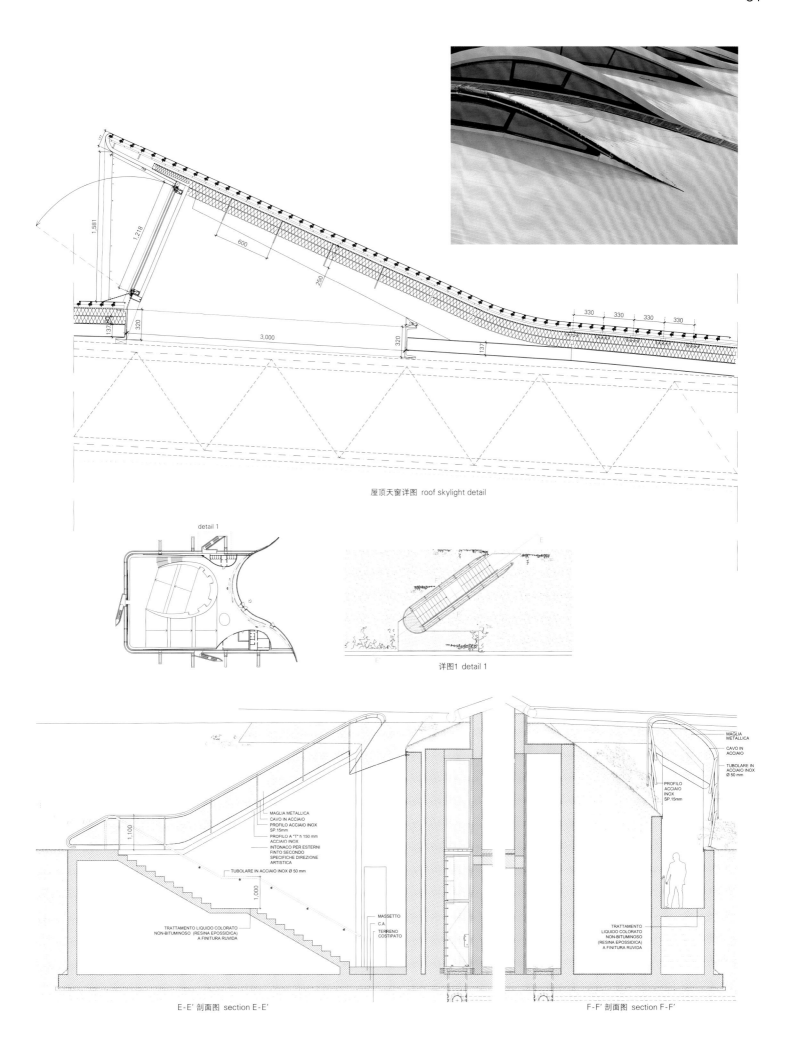

屋顶天窗详图 roof skylight detail

详图1 detail 1

E-E' 剖面图 section E-E'

F-F' 剖面图 section F-F'

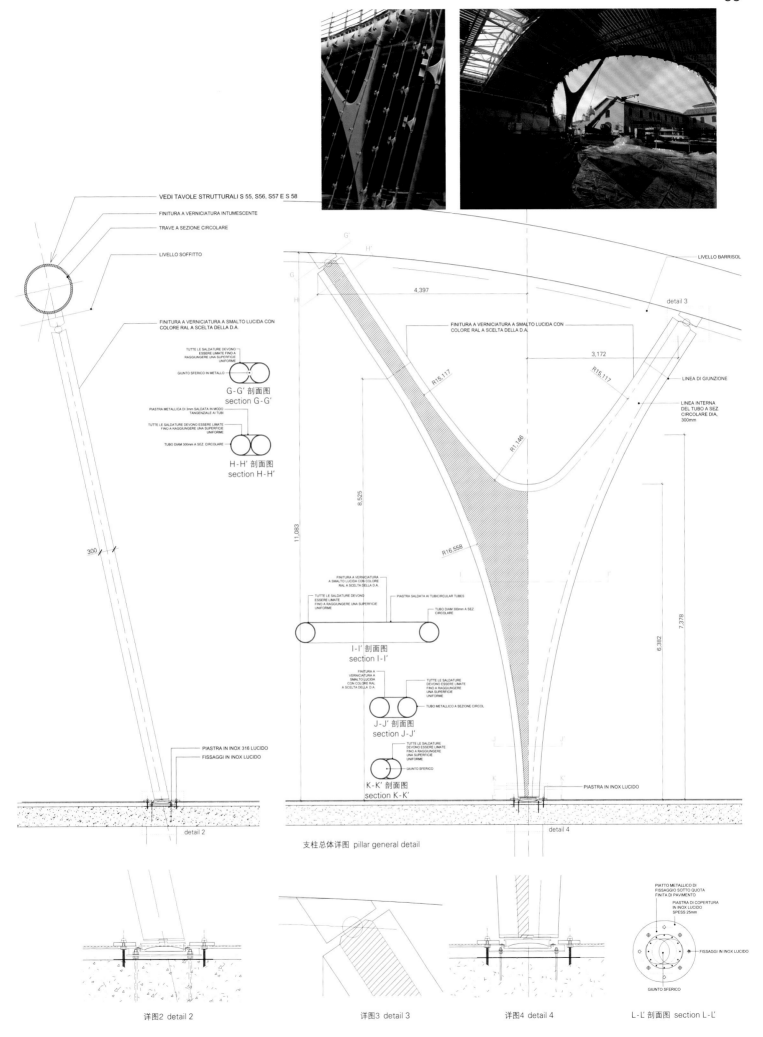

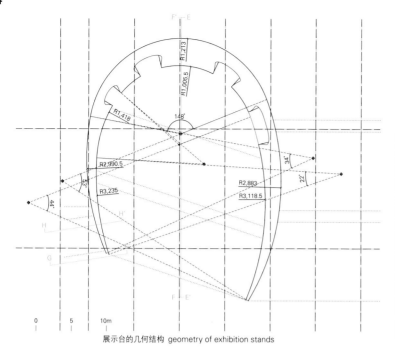

展示台的几何结构 geometry of exhibition stands

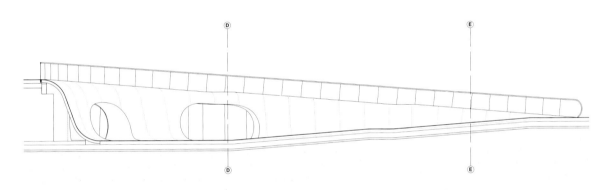

E-E' 剖面图 section E-E'

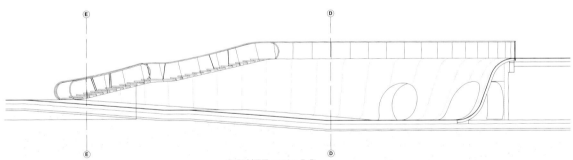

F-F' 剖面图 section F-F'

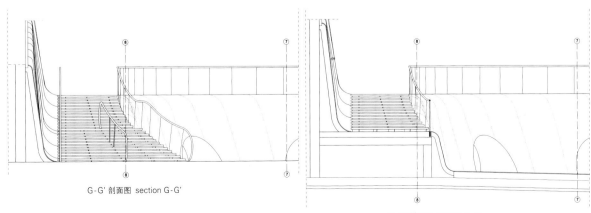

G-G' 剖面图 section G-G'　　　　　　H-H' 剖面图 section H-H'

the aesthetic values of car design. With its 3,300 square meters of double-curved aluminum, the roof is the first application of aluminum in this way on such a large scale. Working together with boat builders whose familiarity with organic sculpted forms and waterproofing made them the ideal partner, and cladding specialists, the form is constructed from aluminum sheets fitted together using a patented tongue and groove system. The bright Modena yellow of the roof is Ferrari's corporate color, as seen on the Ferrari insignia where it forms the backdrop to the prancing horse. It is also the official color of Modena.

Kaplický wanted to create a sensitive dialogue between the two exhibition buildings to show consideration for Ferrari's early home and underscored the importance of the museum as a unified complex made up of several elements. The views out of the new exhibition building dramatically frame the house and workshop, while views from outside the house and workshop immediately reveal the function and content of the new exhibition building. The height of the new exhibition building reaches a maximum of 12 meters – the same height as the house – with its volume expanding below ground level. In addition, the new building gently curves around the house.

The glass facade is curved in plan and tilts at an angle of 12.5 de-

grees. Each pane is supported by pre-tensioned steel cables and is able to withstand 40 tonnes of pressure. The technical specification of these panes and cables means that greater transparency in the facade is achieved with maximum functionality. In the summer months a thermo-sensor activates the windows in the facade and roof, allowing cool air to circulate. With 50% of the internal volume of the main exhibition building being set below ground level, geothermal energy is used to heat and cool the building. It is the first museum building in Italy to use geothermal energy. The building also employs photovoltaic technology and water recycling systems.

Visitors entering the new building have uninterrupted views into the entire exhibition space: a large, open, white room, where the walls and floor transit lightly into one another and are perceived as a single surface. A stretched semi-transparent membrane spreads light evenly, and in combination with the slits running from side to side which allow air to escape and give a ribbed effect, recalls the language of a car's interior. A bookshop and a cafe are situated to one side of the entrance and facilities to the other. Both are painted the same Modena yellow for the roof and take the form of blister-like pods. A gently sloping ramp gradually leads the visitors around the building from the ground floor to the basement level,

with display stands designed by Morgante punctuating the circulation path. These stands lift the cars 45 centimeters so that they can be viewed from different angles and appreciated as works of art rather than objects simply placed in a room. Up to twenty-one cars can be displayed in this open space at any one time. Supplementary exhibition materials are displayed in leather cases located along the perimeter wall. At the bottom of the ramp and directly below the entrance, an audiovisual room forms a permanent part of the exhibition. A flexible teaching space and a conference room with a carved-out opening allowing views up into the entrance area are located next to it.

1 主入口　　　　1. main entrance
2 展览空间　　　2. exhibition space
3 展览室　　　　3. exhibition room
4 会议室　　　　4. meeting room
5 办公室　　　　5. office
6 卫生间　　　　6. toilet
7 档案室　　　　7. archive room

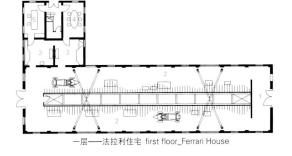

一层——法拉利住宅 first floor_Ferrari House

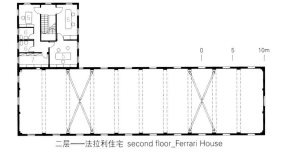

二层——法拉利住宅 second floor_Ferrari House

旧厂房的空间蜕变
Spatial from Temporal
Uptodate in outdated factory

　　对于设计师来说，改造原有建筑显得尤其有趣。之前的境况虽然在空间、构件的使用和新成分的引入方面都产生了极大的限制，它却要求设计师们能够敏锐地发现能够将新的增建结构与过去的建筑相联系的线索，并不时地获得显著的成果。

　　当一座废弃的工厂提供了原有的条件时，这项挑战的本质看起来就相当强烈而且清晰。这一切都建立在已然消失的时间里，为那些想要重新赋予它们生命的设计师们提供了有趣的精神食粮。老旧的支撑结构还坚固地存在着，原有生产设备的痕迹清晰可见，加之原有的工作间的装饰理念所留下的痕迹，这一切都成为现有项目的设计师们的挑战因素。新项目因此提供了一个契机，将过去新用，同时通过当代方式来利用原有旧元素，以此来进行定位。

　　经过这一过程，建筑将成为不同时代的混合体：介于成为旧时的见证（对空间和作品的看法较为老旧），以及新一代为了重新利用和建立作品模型而开拓已有空间的代表之间。

Working with an existing building appears to be particularly interesting for designers. A previous situation often offers significant constraints in terms of space, use of elements or introduction of new parts, but it also requires a watchful eye for important clues as to how to link the new intervention with the past and – sometimes – obtain outstanding results.

The nature of this challenge seemingly becomes quite strong and clear when the previous condition is represented by an abandoned factory. Built in a vanished time, these provide interesting food for thought of the designers asked to bring them new life. The strong presence of the old supporting structure, visible traces of the previous production devices and what is left of the ideas regarding work-space decoration of a former time are amongst the factors that comprise the challenge for the designers of the presented projects. The new project thus becomes an opportunity to determine a position with regard to the past for the new use and – at the same time – an opportunity to design using old elements in a contemporary way.

The building will be – as a result of the process – a temporal hybrid: suspended between being an old witness of past times, likely with an older view of space and work, and being a representative of a new generation, reclaiming the given space for a new use and working model.

建于原马德里屠宰场里的红牛音乐学院_Red Bull Music Academy in the Old Madrid Slaughterhouse/Langarita-Navarro Arquitectos

IMd工程顾问公司_IMd Consulting Engineers/Ector Hoogstad Architecten

红牛公司新总部_Red Bull New Headquarters/Sid Lee Architecture

Tesa 105改建项目_Conversion Tesa 105/Andrés Holguín Torres+David R. Morales Hernández

Zap' Ados/Bang Architectes

米诺——意大利米利亚里诺青年旅社_MiNO, Migliarino Youth Hostel/Antonio Ravalli Architetti

La Fabrique——创新机器_La Fabrique, Creative Machine/TETRARC

不同时代的混合体_Temporal Hybrids/Silvio Carta

不同时代的混合体
新式设计与旧式设计的见证
Temporal Hybrids
The New Design and the Old Witness

Vacant NL——第12届威尼斯建筑双年展上的荷兰馆,由Rietveld景观事务所设计
Vacant NL – Dutch Pavilion at the 12th Venice Architecture Biennale by Rietveld Landscape

尽管时间概念一般被分为三个阶段,即过去、现在和将来,但在调查这一分段方式中,也有一些新想法的线索浮现出来。每个人的行动都被自动地定位在这三个范围之内。该设计(本质上是一种投影行为)像桥梁一样连接了现在和将来发生的行为。现在的行为通常被发生在过去的行为所驱使。因此现在既是过去也是将来,而将来也终将成为过去。

这些无谓的重复如果得到合理使用,能够帮助我们观察人的行为和建筑。支撑这篇文章的思想就是,追寻两个时间"阶段"之间的联系并不如我们所想象的那样锐利、清晰和简单。同样,一种嵌入建筑总是置于另一种能够在某一过去时间找到根据的嵌入建筑中。

谈论一个"新"项目是那么浅显、简单吗?若是这样,相对于"旧"模式,新项目究竟在多大程度上可以说是真正的标新立异呢?本书所提及的项目针对建筑对时间的处理提供了有形的例证,其中包括建筑构件的时间性。

正如有人在这里提及的项目中所观察的,它们显示出并不是每一个设计都仅仅与现在或是过去相关,而是不仅展现出一座建筑与先前的建筑外形相关的方面,同时也能够展现出全新的一面。这两种情况在同一个项目中并存——在设计实现之后——建筑既不是绝对的新,也不是完全的旧。在这里我们所提出的观点是这些建筑占据了两个"时间段"的中间位置。而且,"新建筑"——如这里所观察的——正是简单地位于它们"进化"过程中一个静止的时段。人们可以预测到它们将来会经历其他改造。因此,本文特选的建筑可以被认为是占据一个中间位置,例如介于屠宰场和展示场所之间,又或者是一个仓库和一座滑板公园之间。建筑在其连续的进展中被人们所观察,而这幅特别的图画捕捉到的则是它们不同时代的混合性。

例如,IMd工程顾问公司所提出的重建过程问题近年来在荷兰引起热议。在超级荷兰末期(1989年至2001年之间),这个低地国家越来越多地将注意力转向了原有的、分散的建筑环境。这方面的代表项目就是Vacant NL,是专为2010年第12届威尼斯建筑双年展的荷兰馆而设计的,由荷兰建筑学会(NAI)委任并建造。

荷兰建筑师Ector Hoogstad被一家工程公司(IMd工程顾问公司)委任重建鹿特丹东南部一座老旧的钢铁厂,因而许多问题正处于调查中。一些建筑问题循环出现,包括空间、功能、材料和建筑本身,它们都成为

Although the diffused notion of time is divided into three phases – namely past, present and future – in investigating this apportionment, several clues for new considerations can emerge. Every human act is automatically placed in one of these three categories. The design – in its essence of being a projection – bridges an action in the present to one in the future. The action in the present is often driven by another that occurred in the past. The present is thus also the past and the future, and this latter will become the past as well.

Tautologies such as these – if properly used – can help one observe human actions and, with them, architecture. The idea supporting this text is that the lines between two "phases" of time are not as sharp, clear and easy to trace as one might imagine. By the same token, an architectural intervention is always placed within another intervention which finds its foundation in a time already past.

Is it so obvious or simple to speak about a "new" project? If so, how much of this project is really new, making a clear statement against an "old" mode? The projects presented in this book offer tangible examples of how architecture deals with time, including the time of its elements.

As one observes the projects presented here, it emerges that not every design relates itself to the present or the past only. They all present aspects that one may connect to the previous configuration of the building and – at the same time – also represent something completely new. These two conditions coexist in the same project and – after the realization of the design – the buildings are neither utterly new, nor completely old. The reading proposed here is that the projects occupy a middle position between the two "times". Moreover, the "new buildings" – as has been here observed – are simply caught in a still moment of their "evolution". One may also expect that they will undergo other transformations in the future. Hence, the buildings here featured can be considered as occupying an intermediate position in their existence between being a slaughterhouse and an exhibition space, for instance, or a warehouse and a skate park. The architecture are observed in their continuous fieri (progress), yet captured by this particular picture in their "temporal hybridness".

The IMd Consulting Engineers, for instance, raises the issue of renewal processes, which have become a hot topic in recent years in the Netherlands. After the end of the SuperDutch period (spanning between 1989 and 2001), the lowland country has increasingly turned its attention to the existing and diffused built environment. A representative project in this sense is Vacant NL, made and commissioned by the Netherlands Architecture Institute (NAI) for the Dutch Pavilion to the 12th Venice Architecture Biennale in 2010.

Several questions were thus under investigation as Dutch architect Ector Hoogstad was commissioned by an engineering firm to renew an old steel plant in the southeast of Rotterdam, named after the IMd consulting engineers. The recycling of several architectural questions, including space, function, materials or the building itself is one point of the argumentation, while dealing with the vacancy affecting the Netherlands was a second.

照片提供：Bang Architectes (©Julien Lanoo)

Zap' Ados青年中心和滑板公园，由原有的工业大厅改建而来
Zap' Ados youth center and the skate park, conversion of the existing industrial hall

照片提供：Antonio Ravalli Architetti

旧麻厂改建为米利亚里诺的一处新市中心和青年旅社
Conversion of an old hemp factory into a new city center and youth hostel for the town of Migliarino

争论的第一个焦点，而处理影响荷兰的闲置空间的问题则成为争论的第二个焦点。

在IMd工程顾问公司 (IMd Raad-gevende Ingenieurs) 的新项目中，人们认为原有建筑只是建筑的表皮和外部特征被"用过"了而已。主要的内部空间和总体布局会促使建筑一定按照既定的参数来进行建设。而也有人注意到新办公室已经被构想成一种新的、独立式空间，嵌在原有的工厂里，最终原有的工厂只会成为一个容纳空间和支撑骨架。原有的钢铁厂起到框架作用，同时也成为新的办公室空间的背景。这种背景是有意形成的，因为所有构件都隐藏在多彩的可塑新元素（从垂直封闭墙体到原有的天花板和混凝土地面）之后。建筑之前的外观经过这样构想，让人回想起无法解释的过去，就像是一幅挂在墙上的、描绘了这座建筑多年前模样的巨大图画一样。同时，从外部看，建筑公开表明了它曾经在港口中作为工业建筑的地位。从感知方面看，硬质的工业元素（仍旧是地板、金属天花板、原有的砌体墙和巨大的天窗洞口）材料的粗糙性与新介入元素（作为隔墙的半透明塑料挡板、内部某些表面的黄色光滑镉染料和诸多"城市家具"，如花瓶里的植物或是在普通休闲空间仿草坪式的绿色地毯）的光泽度与锐利性相结合，两者的对比赋予新IMd办公室以启发性的独特氛围。此外，正如讨论中所提到的，该项目标志着新建筑处理原有建筑条件的重要转折点。

在IMd工程顾问公司将新元素融入到建筑原有特色当中的意图仍旧有迹可循时，红牛公司总部中建筑原有特色已经不可见了。事实上，Sid Lee建筑公司似乎已经向着全然不同的方向前进。阿姆斯特丹北部的原造船厂中新建的介入物预示着打破原有规则的新思想，避免与任何先决条件进行对话。根据几种观念——材料的使用、色彩的选择、空间的层级或室内布局——的启发，新项目清晰而公开地表示与环境截然不同。红牛公司新总部不仅仅是位于原有建筑的内部，而更像是插入或是应用在了内部空间里。事实上，原有建筑所具有的工业特征和外观为新插入的建筑提供了方便的"白画布"。在这个意义上，这座砖质建筑由于其自身的空间性和功能性已经不再是一座综合建筑物，而成为一个外壳——或说一个容器。一旦它完美地转变为框架，就允许建筑师们在显著的位置建立新的可被感知的元素。而新建的关系不像是对话或者"转换"，而更像在一个空间里按照新的顺序将各种构件叠加在一起。这一选择当

In the new project for the IMd consulting engineers (IMd Raad-gevende Ingenieurs), one apprehends that the existing building has been "used" only for its shell and outer features. The main internal space and gross layout have certainly driven the project with given parameters, yet one notices that the new offices have been conceived as a detached, new insertion within the former plant, which eventually remains only as a container and supporting skeleton. The steel plant plays the role of framework and – at the same time – background for the new office spaces. This backgrounding is intended, as everything is seen behind the colorful, plastic new elements, from the vertical closed walls, to the existing ceiling and the concrete floor. Conceived as such, the "former" appearance of the building remains as a reminder of an unexplained past, in a way similar to a large painting hung on a wall depicting the same building in years past. Meanwhile, from the outside, the building openly declares its previous status of an industrial presence in the harbor. In perceptional terms, the contrast obtained by combining the roughness of the material belonging to hard industrial presences (again, floor, metal ceiling, the original masonry of the walls and the large skylight openings) and the glossiness and sharpness of the newly inserted elements (plastic translucent sheeting as partitions, slick cadmium yellow for certain internal surfaces and a variety of "urban furniture" such as in-vase plants, or green carpet simulating grass in the common recreational spaces), confers upon the new IMd offices a stimulating and distinctive atmosphere. Moreover, as concerns this discussion, this project marks an important turning point in the way a new architecture can deal with an existing situation.

While in the IMd consulting engineers, the intention to relate the new intervention to the existing features of the building can still be traced, in the Red Bull headquarters the latter is no longer visible. The architectural firm Sid Lee Architecture, in fact, seems to have headed in a completely different direction. The new interventions in the former shipping yard in the north of Amsterdam betoken the idea of breaking the "rules" of the existing, and avoiding any sort of dialogue with the pre-condition. Under the light cast by several points of view – the use of materials, choice of colors, spatial hierarchies or internal distribution – the new project is clearly and openly making a break with the context. Rather than being situated within the existing building, the new Red Bull headquarters seem more to be inserted into or applied to the internal space. The existing building, in fact, due to its industrial features and appearance, offers a convenient "white canvas" for the new "insertion". To this extent, the brick building stops being a complex, with its own spatiality and functions, and becomes a shell – a container. Once ideally turned into a framework, the brick building allows architects to build new elements to be perceived in the foreground. The newly established relationship is not that of a dialogue or "exchange"; it is more of a superimposition of a new order of subjects onto a space. This choice is of course related to the new user of the building, the company Red Bull, which has been recognized for its outside-the-box way of

IMd工程顾问公司，前身是一家钢铁厂
The IMd consulting engineers, formerly a steel plant

红牛公司在阿姆斯特丹的总部，位于一家历史遗留下来的造船厂
Red Bull Amsterdam's headquarters landed in an old heritage shipbuilding factory

然与建筑的新用户——以创造性的思考和产品展示方式而著称的红牛公司有关，那么建筑师们对原有环境采用出乎意料的——甚至从某种程度上说是有些惊人的处理方式也就不足为奇了。其内部空间的创建和装饰的广泛应用似乎与这幅独特的草图无关，也并不是它的一部分，除非人们有意以惊诧的眼光去观察。同样，所有的元素共同作用促成了一幅连续不断的透视图。

意大利一座510m²的青年旅社项目建于原有工业建筑内，由意大利的Antonio Ravalli Architetti设计。该项目意在将原有环境（线条呈直角，且僵硬）与以圆形和圆柱形为基础的新介入建筑进行明显的对比。在空间和形状方面，新房间被设想为寄生虫，即寄养在宿主内的不同性质的构件。建筑师们暗示说，房间"就像是室内营帐，独立的小房间外面包裹着轻质覆层"。建筑师的描述中显现出解读该项目的一些线索。首先项目的目的不是要将新的工程植入原有建筑所提供的"空白"空间。他们选择不把原有场地的空间当做新项目的一种基本元素，而是当做已经存在的背景，就像用来野营的森林。事实上，房间——置于排列紧密的七根柱子中间——根据一种新的（空间的）逻辑建于原有建筑空间之内，看上去只是简单地叠加在原有的一家工厂之上。如果这些圆柱形状的新空间的原则被看做是自主的，而新的结构也与原有的相关，那么我们就可以简单地断言这些圆柱是在空间内自由排列的。第二点发现是：这些圆柱被建筑师们看成是自主的元素。这一构想表明建筑师们将原有建筑与新的介入建筑区别看待的思想，因此从根本上证明了对话这一术语——如果这一术语在此种情况下所用合理——正是基于新旧建筑的某种不相容性。更进一步思考这种构想，这里也潜藏着兼容性的问题。与原有空间已经存在的正交的和方形既定空间相比，为什么新房间选择以圆形为基础的形状（如圆柱）作为新设计的固定点或是起始点就被认为是有着独特的本质呢？又是从哪一点来说方形空间就真的不能与以圆形为基础建造的空间相容，两者之间就不能达成对话呢？换言之，为什么这些新建的"小房间"在处理方形空白空间时需要"自主"呢？况且，如果新的介入建筑不能与已存的建筑交流，换句话说，就是两者不能通过一种相容的、非自主的方式建立对话，那么这种方式又是因何而产生呢？支撑青年旅社新房间的七根圆柱与其说是建筑环境，是不是更像是一种构图结构呢？

thinking about and presenting products. It is thus not surprising that the approach of the architects regarding the existing context would be unexpected and – in some way – startling. The internal spatialities created and the extensive use of decoration seem not to be related to any part of a unique drawing, unless they are observed under the lens of intentional surprise and amazement. By the same token, all elements work together, contributing to a continuous scenography.

The project in Migliarino, a 510m² youth hostel inside a former industrial building by Antonio Ravalli Architetti in Ferrara (Italy), shows an intention to establish a clear contrast between the context – square and rigid in its lines – and the new intervention, based on circular and cylindrical shapes. In terms of space and shape, the new rooms are conceived as parasites, intended as elements of a different nature inserted in a hosting body. The rooms are "like indoor camping, autonomous cells enfolded in a light wrapping", suggest the architects. A few clues for how to read the project emerge from the architects' description. The first is the intention not to root the new project into the "empty" space offered by the existing building. They have chosen not to consider the onsite space as a fundamental element of the new project, but as an already-there setting, figuratively a forest in which to camp. As a matter of fact, the rooms – contained in seven squeezed cylinders – are placed in the space according to a new (spatial) logic which appears to be simply superimposed onto one of the former factories. If the new spatial tenet of the cylinders is to be considered autonomous, and if the new configuration is related to the existing, it could easily be asserted that they are placed freely in the space. The second observation is that the cylinders have been considered by the architects to be autonomous elements. This conception indicates an idea of considering the existing building and the new intervention in separation, thus basically demonstrating that the dialogue – if the term is appropriate in this case – has been based on a sort of incompatibility between the existing and the new. To take this speculation further, there lurks the question of compatibility. Taking the choice of circle-based shapes (such as the cylinders) for the new rooms as a fixed or starting point for the new design, why are they to be considered as having a distinct nature with respect to the existing orthogonal and square-based given space? At what point are the square-based spaces not really compatible with those constructed based on the circle, in the sense that a dialogue cannot be established between them? In other words, why do the "cells" need to become "autonomous" while they are dealing with the squared empty space? Moreover, if the new insertion is unable to communicate with the existing architecture, namely by establishing a dialogue in a compatible and un-autonomous way, what could the reason for this approach be? Can the seven cylinders hosting the new rooms of the hostel be closer to a scenographic configuration than an architectural setting?

The dialogue between existing condition and new intervention reaches an interesting point in Bang Architectes' new skate park in

照片提供：David R. Morales Hernández (©Andrea Pertoldeo)

Tesa 105，改建自威尼斯兵船厂北部的一座16世纪工业仓库
Tesa 105, conversion of a 16th century industrial warehouse in the northern section of the Venice Arsenale

在Bang建筑师事务所设计的新圣皮埃尔滑板公园项目中，原有条件和新建介入建筑之间的对话达成了一种有趣的共识。公园前身是法国加来的工业区。原有的仓库——之前是一家烤花生工厂——并没有被看成是能够在其中进行填充的外壳或骨架，也没被看成是用来建立对比关系的既定条件。在建造新项目中，法国设计师利用了工业建筑的显著特色。新建筑最后完全成为一个全新的实体，而不是涂了新染料的旧框架。在加来的新滑板公园中，过去的痕迹在建筑的主要形状、空间组织和内部承重结构方面仍然清晰可见。然而，这些痕迹都是新建筑的组成部分，这座建筑似乎终于将这些元素融入到新的结构特征里。在这种意义上，新项目围绕着原有建筑所提供的设计"线索"进行构思，而这条线索也的确适合于这个仓库。它的工业特色没有被忽视或者弃置一旁，也没有加以掩饰或隐藏。例如，设计师们创建了一个新的橙色金属网，像表皮一样包覆起整个仓库。这本身并没有什么有趣的特色，而当它与穿越建筑内部的橙色金属篱笆（将滑板场地与纵向走廊以及封闭房间分隔开来）相联系，再配以走廊的橙色地板，橙色的表皮所表露的就不再仅仅是外表面。"橙色"使这些元素都成为同一体系的组成部分，表明它们并不是孤立的成分，也不是彼此独立。它们作为同一个建筑姿态或创作意图的一部分而发挥作用。原有的仓库不再作为一个外壳，而是一层薄纱，新建筑如线一般在此交织。从这一点看，其他的建筑也可以看成是有同样意图的产物。使用穿孔金属网格代替不透明的金属板产生了一种由外至内的极大透明感，将原有仓库的表面进行了变形。尽管形状和主要轮廓仍旧相同，但不透明性的变化却显而易见，这也意味着更强的采光和视觉效果将使业主和路人对建筑产生更大兴趣。运河边上的这片工业区的天际线仍然可见，而新项目的创新性也相当显著。另外，滑板公园的内部空间非常清晰地显示了裸露的（修复的）结构元素和建筑的原有部分，以原来支撑屋顶的桁架形成的拱顶为特色。这样一来，原有的空间性得以留存，只是以一种新的面貌呈现。

还是在Matadero，音乐仓库（the Nave de Música）——举办每年一度的红牛音乐学院音乐节的当代创造性空间——由马德里Langarita-Navarro Arquitectos设计完成。项目从Matadero的老仓库外部找不到任何线索，而是通过在内部创造出一个完全不同且出人意料的空间来与原有仓库保持距离。原有的环境与新项目的关系正是基于将旧建筑看做框架

St. Pierre, the former industrial area of Calais, France. The existing warehouse – a former roasted peanut factory – is not considered a shell or a skeleton to be filled in, or a given condition with which to establish relationships of contrast. The French designers have taken advantage of the industrial building's distinctive features in building the new project. Rather than being an old framework with a new painting, the final result is truly a completely new entity. In the new Calais skate park the traces of the past are actually still visible, in the main shape of the building, in its spatial organization and in the internal bearing structure. However, those traces are part of a "new" building, which seems to eventually have integrated these elements into its new constitutive features. The project is, in this sense, conceived around design "hints" given by the existing building, and it is literally suited to the warehouse. The industrial features are not left aside or ignored, nor are they dissimulated or hidden. The designers, for instance, created a new orange metal mesh cladding the entire warehouse as an external skin. This fact does not represent any interesting feature per se, but when related to the orange metal fence which crosses the inside (separating the skate rink from the longitudinal corridor and the closed rooms), along with the orange floor of the corridor, the orange skin emerges as not merely an outer surface. The "orange" makes the elements part of the same system, declaring that they are not isolated components, nor independent from one another. They work as part of the same gesture and compositional intention. The existing warehouse is no longer a shell, but a tissue, through which the thread of the new project has been sewn. From this point, other architectural choices can be observed as consequences of the same intention. The use of a perforated metal mesh instead of opaque metal paneling results in a significant outside-inside transparency, which tweaks the appearance of the former warehouse. The shape and the main silhouette remain the same, while a variation in opaqueness is apparent, and also means more lighting and visual effects confer more interest upon the building for both users and passers-by. The skyline of the industrial area along the canal is still recognizable, while the novelty of the new project is also quite apparent. Moreover, the space inside the skate park shows rather clearly the bare (restored) structural elements and the original section of the building, featuring a vault created by the original girder supporting the roof. The original spatiality has been – in this sense – maintained, although with a new appearance.

Within the same context of the Matadero, the Nave de Música (music warehouse) – a temporary creative space hosting the annual Red Bull Music Academy (RBMA) festival – has been designed by Madrid-based Langarita-Navarro Arquitectos. This project takes distance from the existing warehouse by creating a completely different and unexpected "world" in the inside, with no clues from the outside of the Matadero's old warehouse. The relationship between the existing condition and new project is based on the idea of considering the old building as a framework, in which the new program can fit comfortably. The several new parts of the project

马德里Matadero红牛音乐学院，位于20世纪初的一座工业仓库综合建筑中
Red Bull Music Academy in Matadero Madrid, located in an early 20th century industrial warehouse complex

而新项目可以与之充分相容的思想。正如建筑师所描述的，项目的几个新部分凸显了"不改变仓库本身，而是精确地保留其他建筑介入之前的原状"的意图。新的部分建立了新的内部景观，同时将外部景色与原有建筑的内部看作连续的背景。

由法国TETRARC建筑师事务所设计的新表演艺术综合建筑"Fabrique"，代表了将新项目与原有环境相联系的另一种有趣尝试。法国西部城市南特的Île仍能追寻到辉煌的造船工业的痕迹，该工业于1760年左右兴起，即从Julien Dubigeon在南特市建立了这座经年累月的造船厂开始。1987年起，这种一度繁荣的造船工业渐渐衰落，仓库和设备也逐渐被弃用。在以"反思时期"（见http://www.iledenantes.com）为特点的1998年之后，Île经历了重要的城市复兴工程，许多设计师、顾问和当地居民群策群力，试图找到这一弃用的工业区和造船区得以重新居住的方法。在这一框架之内，TETRARC建筑师事务所设计了新音乐厅。新介入建筑的关联性明显地体现在项目的工作室方面，工作室叠加在旧防空洞上。在这一项目中原有结构和新体量的结合具有显著不同之处：厚重的混凝土块成为新设计结构的矮墙，一个悬置的空间——像是一个底层架空柱——被留在两者之间，更加凸显同一座建筑的过去与现在环境之间的差异性关系。

另一个用外壳式手法处理的建筑以Estudio N设计的威尼斯兵船厂里的Tesa 105为代表。这座16世纪的建筑是较大的威尼斯兵船厂的一部分（威尼斯兵船厂于1104年左右开始建造），曾经致力于军械和兵船的制造，现在已经被改建为一处充满活力的新活动中心，它的一层是酒吧和聚会场所，二层是办公室和设施区。威尼斯的历史意义本身足以保证原有建筑外部能够轻易恢复，甚至不受任何影响。它的内部因此成为实施新规划和全新设计的场所。像威尼斯一样历史悠久而保守的建筑背景以其看似不变的外壳与差异极明显的内部结构所产生的巨大反差为特征。在这里我们可以谈及两个作品：一个是自16世纪原样保留至今的原有建筑，一个是新的设计（例如，基于当代需求、要求、品位、思想和构想的设计）。描述新旧项目的关系可以借用"公然的裂缝"一词来描述。原有建筑谨慎地与新的介入建筑分离开来。事实上，新设计似乎并不能与已经存在的建筑进行交流。原有的建筑被Estudio N处理成一个布置新规划的背景，在最近的建筑历史时期，它是新介入建筑无声的古老见证者。

show the clear design intention of "not modifying the warehouse itself, but rather leaving it exactly as it was before the intervention" as the architects described. The new parts establish a new internal landscape, while considering both the outside views and the inner part of the old building as a continuous background.
The new "Fabrique", a performing arts complex designed by the French architects of TETRARC, represents another interesting attempt to link a new program to an old context. The Île de Nantes still carries the traces of a significant naval industry, which began around 1760 when Julien Dubigeon founded a long-standing naval shipyard in the city of Nantes, in the west of France. From 1987, the once-flourishing industry ground to a halt, leaving the warehouses and facilities in a slow and continuous process of abandonment. After what has been characterized as a "A Time for Reflection" (see http://www.iledenantes.com) in 1998, the Île underwent a significant urban renewal project, in which several designers, consultants and local residents worked together to understand how the unutilized industrial and shipyard area could be rehabilitated. Within this framework, TETRARC designed the new music halls. The relevance of their intervention is clearly apparent in the aspect of the project dedicated to the studios, superimposed on an old air-raid shelter. In this project, the combination of the existing structure and the new volume is marked by severe distinction: the massive concrete block becomes the podium for the new design, and a suspension space – like a pilotis – is left between the two, emphasizing even more the distant relationship between the past and present "conditions" of the same building. Another example of the "in-shell" approach is represented by the Tesa 105 in the Arsenale area of Venice, designed by Estudio N. The sixteenth-century building is part of the larger complex Arsenale di Venezia (the Venetian Arsenale, whose construction began around 1104), formerly devoted to an armory and shipyard, and now converted into a lively new activities center, with bars and meeting spots at the ground level and offices and facilities on the second above-ground level created. The historical significance of the city of Venice is enough in itself to justify the outside of such existing buildings being simply restored, almost untouched. The inside thus becomes the realm in which new programs can be hosted and which allows for a brand new design. Such historical and conservative contexts as Venice are characterized by a contrast between a seemingly unchanged outer shell and an inside content which makes a clean break. We can speak here of two projects: the existing from the sixteenth century preserved as it was, and a new design (i.e., based on contemporary needs, requirements, tastes, ideas and concepts). The relationship of the old and new is here described by means of a declared hiatus. The existing building is reverentially separated from the new intervention. The new design seems, in fact, unable to communicate with the presence already there. The existing building has been treated by Estudio N as a background on which to place the new program, an old witness silently observing a new intervention, in this most recent phase of the building's history. *Silvio Carta*

这个项目在很多方面借鉴了俄罗斯套娃的形式。不仅在最表面和物理层面上,在时序感受上亦是次第形成。这个项目的最初环境为这种情形建立了一种有利的背景。

紧急项目,2011年马德里红牛音乐学院

红牛音乐学院(RBMA)的音乐节是一年一度举办的流动性音乐节。在过去的十四年里,音乐节在世界各地不同的城市举办,迎来了六十位预选的国际参与者,并有音乐家、制作人和电台DJ支持,因此给他们提供了体验、交流世界音乐知识和理念的机会。2011年的红牛音乐学院音乐节原本定在东京举行,但由于地震的灾难性影响,不得已而改变地点。马德里市接手了该项目,且只有五个月的时间来筹备。一个名为马德里Matadero的创意空间位于20世纪初期的工业仓库综合建筑里,它被指定为举办音乐节的新场地。

一个中期项目,马德里Matadero的音乐仓库

红牛音乐学院开展音乐仓库的规划工作,该仓库是专门进行音频创作和研究的空间。以现有的装置为起点,再加上其实验性特质,这个建造项目被处理成基于适应性和可逆性准则的临时性结构,从而易于以后进行全部或者部分改装。

在这些环境和此种紧急情形下,项目以满足音乐节的精准技术和声学要求的基础设施为开端。

方案在以下五点指导原则下展开。

1. 工期和预算

设计必须明确地服从一些非常严格的工期要求和预算上的考虑。建设工期必须在两个月内完成,并且只能使用轻质结构的方案,在标准化和适应性之间寻求一种平衡。

2. 尊重仓库现状

Matadero15号仓库是一个带有金属结构的开放性空间,其外部是砖立面。该金属结构的面积约为4700m²,直接对外部开放。该项目强调的原则之一是不改变仓库本身,而是要保留其改建之前的状态。

3. 项目要求

项目清晰地建立了一种布局,该布局分成四个区域:办公室、音乐家工作室、录音工作室和用于会议、广播和休息的区域。选定的空间和结构体系可以满足日后举办大型事件的空间重组要求。

4. 声学要求

每一处区域都获得了一种与用途一致的特定逻辑,这使得独一无二地解决各自的声学需求成为可能。各种各样的解决办法包括录音工作室的重型墙体、会议室布质圆屋顶的吸音表面、不平行排布的亭子的结构和几何独立性。

5. 暂时性要求

该项目设计得易于拆除,甚至"最笨重"的构件都被设计为可复原的,以方便今后大型事件的循环使用。这方面的例子还包括使用沙袋来构建录音工作室的墙体以及日后可以在Matadero其他区域或其他城市进行移植的盆栽植物。

因此,项目以不完整的城市结构形式,在这间仓库室内逐渐展开,而城市结构中邻近性与独立性之间的可变关系、既存状态和性能,为当地社区居民提供了意想不到的舞台。

Red Bull Music Academy in the Old Madrid Slaughterhouse

In many ways this project shares the logic of a Russian matryoshka doll, not only in the most literal and physical sense, but also in a temporal sense in which one actually originates within the other. The initial circumstances of this project established a favorable backdrop for this condition.

An emergency project, the Red Bull Music Academy Madrid 2011

The music festival of the Red Bull Music Academy (RBMA) is a nomadic annual music festival. For the last 14 years, this event has been held in different world cities, welcoming the sixty pre-selected international participants and surrounding them with musicians, producers and DJs, thereby giving them the opportunity to experiment with and exchange knowledge and ideas about the world of music. The 2011 edition of RBMA was going to be held in Tokyo, but given the devastating effects of the earthquake, the location had to be changed. With only five months to plan, the city of Madrid took over. The creative space known as Matadero Madrid which is located in an early 20th century industrial warehouse complex was designated as the event's new location.

建于原马德里屠宰场里的红牛音乐学院
Langarita-Navarro Arquitectos

项目名称: Red Bull Music Academy
地点: Matadero Madrid, Spain
建筑师: María Langarita, Víctor Navarro
合作者: Juan Palencia, Gonzalo Gutierrez, Tonia Papanikolau, Paula García-Masedo
勘测员: Javier Reñones
结构工程师: Mecanismo S.L.
机械工程师: Úrculo ingenieros
声学工程师: Imar Sanmartí Acousthink S.L.
灯光结构: Arquiges y Cuatro50
景观建筑师: Jerónimo Hagerman
用地面积: 4,700m²
建筑面积: 4,700m²
总楼面积: 1,315m²
设计时间: 2011.6
施工周期: 9 weeks
竣工时间: 2011.10.15
摄影师: ©Luis Diaz Diaz (courtesy of the architect) – p.49, p.54, p.57, p.58~59
©Miguel de Guzman – p.50~51, p.53, p.59 right

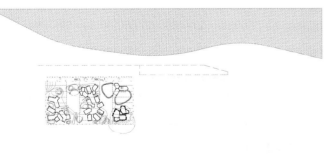

A medium-term project, the Music Warehouse in Matadero Madrid

The RBMA launched the programming for the music warehouse, a space specifically dedicated to audio creation and research. Using the existing installation as a starting point and given its experimental character, the construction project was approached as a temporary structure based on the criteria of adaptability and reversibility that would make it easy to completely or partially reconfigure over time.

Under these circumstances and in an emergency situation, the work began on an infrastructure capable of meeting the precise technical and acoustic needs of the event.

The proposal was developed based on five guidelines.

1. Deadlines and budget

The design had to specifically comply with some very tight deadlines and budgetary concerns. The construction had to be completed in less than two months, implementing solutions that would require only light construction and seeking a balance between standardization and adaptability.

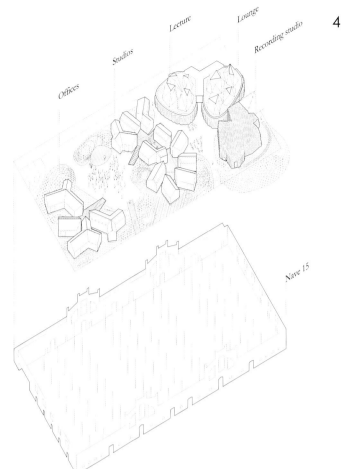

2. Regarding the warehouse

Warehouse 15 of the Matadero is an open space comprised of a metallic structure with a brick facade. This structure which measures about 4,700 m², opens directly to the outside. One of the criteria taken into account for this project was that of not modifying the warehouse itself, but rather leaving it exactly as it was before the intervention.

3. Program requirements

The program's organization clearly establishes a specific configuration that is grouped into four areas: offices, studios for musicians, recording studios and an area used for conferences, radio and lounge. The chosen spatial and constructive systems would allow for the reconfiguration of these spaces for future events.

4. Acoustics

Each of the areas acquired a specific logic that corresponded with its usage, thereby making it possible to uniquely resolve its acoustic needs. Some heterogeneous solutions included the massive walls in the recording studios, the absorbent surfaces of the cloth domes in the conference room and the structural and geometric independence of the nonparallel pavilions.

5. Temporariness

It was designed to be dismantled. Even the "heaviest" actions were designed to be reversible and to allow for their easy recycling for future events. Examples of this included the use of sandbags to make up the walls of the recording studios and potted plants that could later be transplanted in other areas of the Matadero or the city.

As a result, the project unfolded in the warehouse's interior in the form of a fragmented urban structure in which the variable relationship between proximity and independence, and preexistence and performance could offer unexpected stages to its community's inhabitants. Langarita-Navarro Arquitectos

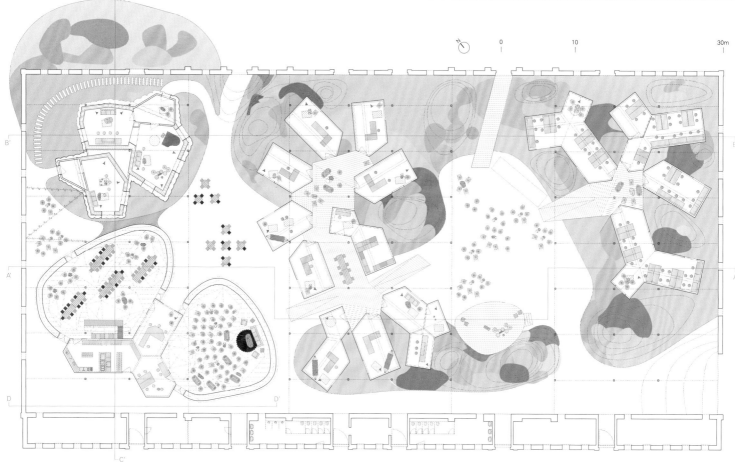

A-A' 剖面图 section A-A'

B-B' 剖面图 section B-B'

C-C' 剖面图 section C-C'

D-D' 剖面图 section D-D'

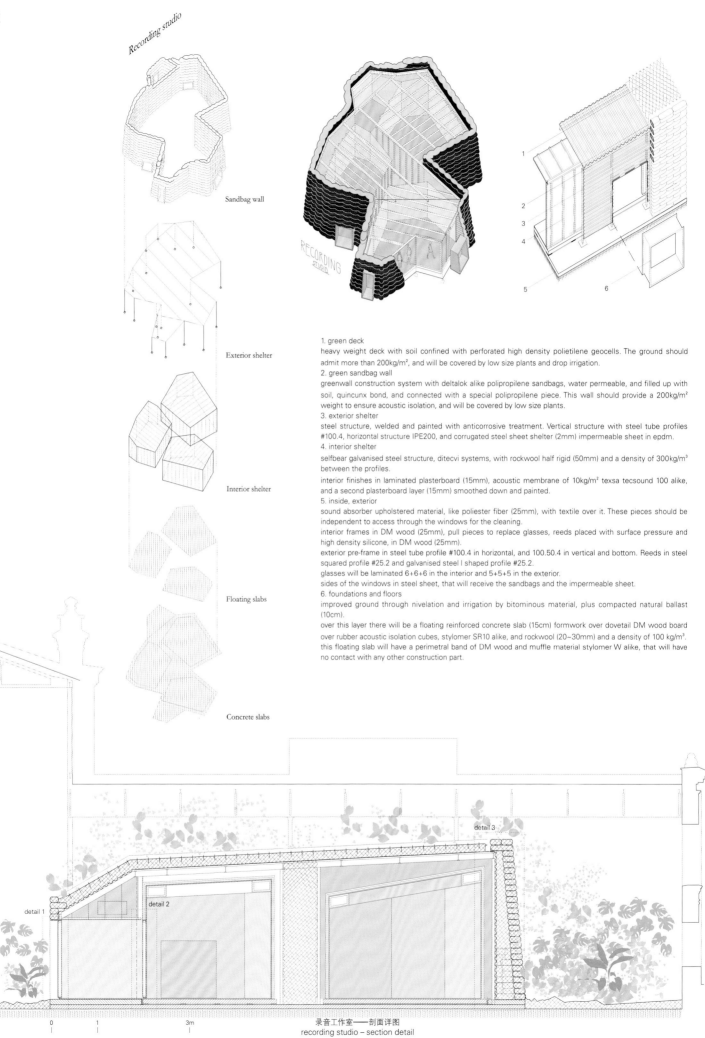

1. steel profile
2. plasterboard ceiling
3. isolation
4. plasterboard
5. wooden floor
6. rubber anti-vibrator
7. ground bags
8. ground

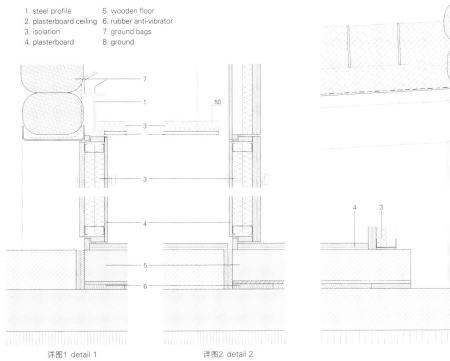

详图1 detail 1   详图2 detail 2   详图3 detail 3

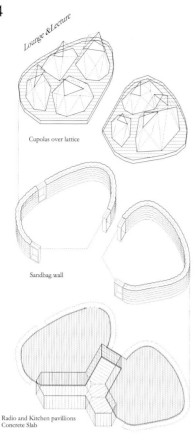

Lounge & Lecture

Cupolas over lattice

Sandbag wall

Radio and Kitchen pavillions
Concrete Slab

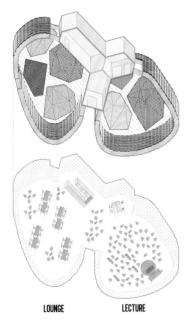

LOUNGE   LECTURE

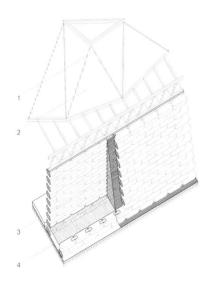

1. cupolas structure
cupolas: double steel tube profile 100.50.4 ring hanging from the preexistent structure by a steel wire.
sides, welded steel tube profile T 25mm.
perimetral lattice by steel tube profiles 100.50.4 in the superior and inferior levels, and diagonals in 50.50.4 supported by the green sandbag wall.
horizontal substructure in steel tube profile 50.50.4 every 60cm. between the perimetral ring and the lower perimetral rings of the cupolas.
both lower rings will also have a folded galvanised steel sheet 70.200.5.

2. cupola finishes
plasterboard (13mm) and rockwool (50mm, 40kg/m³) over painted steel structure.
interior finishes, through capitoné polyester fiber (25mm) upholstered with acoustic absorbent material.
horizontal finishes: cell polycarbonate board (40mm) over painted steel structure.

3. green sandbag wall
green sandbag wall construction system with deltalok alike polipropilene sandbags, water permeable, and filled up with soil, quincunx bond, and connected with a special polipropilene piece. this wall should provide a 200kg/m² weight to ensure acoustic isolation, and will be covered by low size plants.

4. foundations and floors
improved ground through nivelation and irrigation by bituminous material, plus compacted natural ballast (10cm).
over this layer there will be a settled green sandbag wall, and the sanitary slab under the wooden pavilions to host the services.
at the lounge and lecture area there will be a floating reinforced concrete slab (15cm).

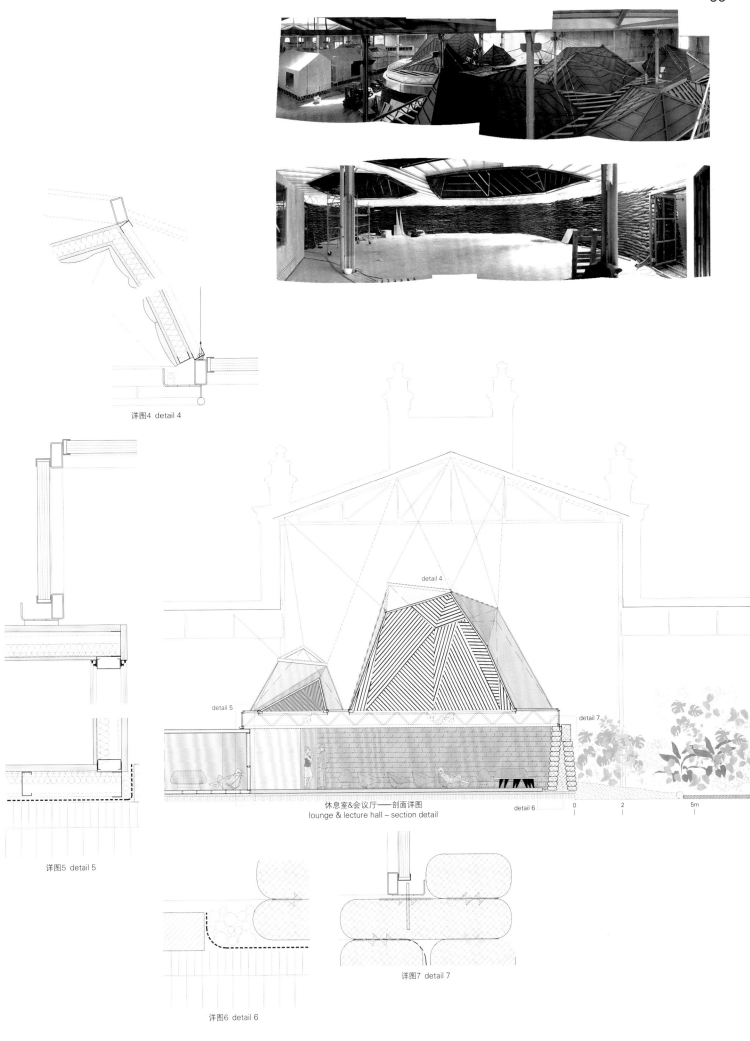

详图4 detail 4

详图5 detail 5

休息室&会议厅——剖面详图
lounge & lecture hall – section detail

detail 4
detail 5
detail 6
detail 7

0 1 2 5m

详图6 detail 6

详图7 detail 7

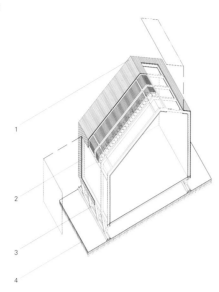

1. wood pavilions (lecture hall, bedroom studios and office studios)
general: vertical and horizontal structure in galvanised steel profiles, structural system ditecvi. Exterior finish in laminated chudoply 22m. wood boards, rockwool (50mm, 70kg/m³), and interior in plasterboard (15mm) plus specific finish in certain zones.
kitchen finish: 10x20cm. vertical white ceramic tiles.
interior acoustic finish: sticked acoustic membrane texsa techsound100 and laminated plasterboard (15mm) smoothed and painted.
bedroom studios and office studios finishes: smoothed and painted
floors in treated laminated wood boards.
2. doors and windows
doors defined as P01 made by the ditectvi system, galvanised steel profile structure, and steel frame with acoustic material in the interior, double seal for pressure closing. The door will have in the inside double rockwool board (50mm, 70kg/m³) and will be finished with a chudoply laminated wood (22mm) outside and plasterboard (15mm) inside.
the vertical windows V01 will have double laminated glass 4+4 within a galvanised steel perimetral l-shaped profile L25.2 and high density silicone.
the upper windows V02 will have double laminated glass 4+4 within a galvanised steel perimetral l-shaped profile L25.2 and high density silicone, supported over the ditectvi system galvanised steel profiles.
3. foundations over supporting lattice
understructure over the ground with a galvanised steel profile structure, ditectvi system alike, with a 20cm support and a total height of 60cm. These lattice will come in all the perimeter and the projection of the top side of every studio.
4. ground preparation
improved ground through nivelation and irrigation by bitominous material, plus compacted natural ballast (10cm).

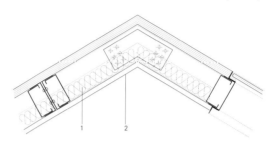

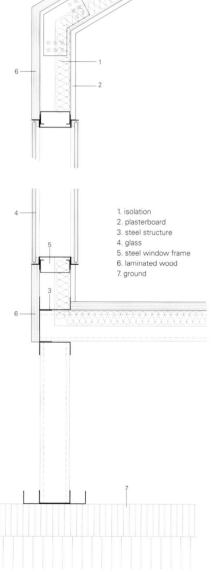

1. isolation
2. plasterboard
3. steel structure
4. glass
5. steel window frame
6. laminated wood
7. ground

一层——工作室
first floor_studio

屋顶——工作室
roof_studio

工作室剖面详图 studio section detail

工作室立面
studio elevation

Offices / Studios

1 Wooden exterior shelter
2 Metalic structure
3 Floating slab
4 Supporting lattice

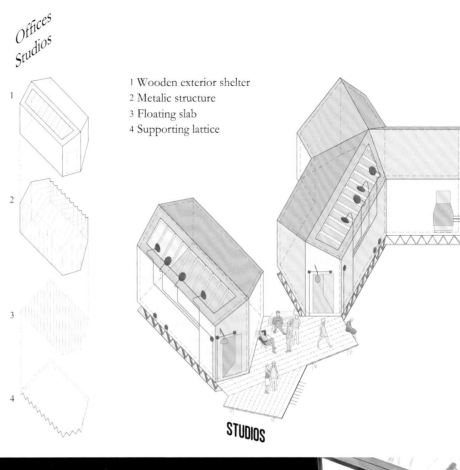

STUDIOS

IMd工程顾问公司
Ector Hoogstad Architects

钢铁厂成为"工程师乐园"

IMd工程顾问公司的新建筑最近由市议员Jeanette Baljeu在鹿特丹宣布开放。IMd没有选择普通的工作环境,而是以一种非正统的方式将一座旧钢铁厂转化成一处"工程师乐园",鹿特丹Ector Hoogstad Architects (EHA)的建筑师Joost Ector这样形容它。

回收利用是当今荷兰的一个大问题。很多建筑都是空置的,等待翻新或重新安置,包括质量上乘的建筑,它们默默等待着可以发现其潜质的人。这家位于鹿特丹Piekstraat的钢铁厂也是如此;它并不是设置办公室的绝佳位置,但是它位置独特,可以俯瞰马斯河的景色。这座建筑吸引IMd的是它由雄伟的钢结构主导的开阔空间。

Ector Hoogstad Architects (EHA)已经与IMd共同合作过许多项目。IMd也参与到EHA在一所旧学校建筑里设计自己办公室的工程中。这次最近的合作也启发了IMd的所有者Remko Wiltjer和Pim Peters为他们的公司寻找独特的建筑。他们不仅看到了自己公司的优势,同时也意识到一座极具特点的建筑物将有助于IMd更加清晰地将自己定位为荷兰领先的设计工程公司之一。他们与新工业开发商一起,发现了这个旧钢铁厂房。

对现有建筑外墙的改造很快被证实在技术和成本两个方面都不切实际。最后,选中的策略决定所有的工作区域都安置在上下两层的空调区,与封闭的端墙相邻。从这里,他们可以向后望见大厅,那里建有带会议区的展馆,其中的人行天桥和不同种类的楼梯将各展馆连接起来。大厅本身就形成了微弱的空气循环空间,成为理想的非正式咨询、讲座、展览以及午餐的场所。原本封闭的立面上设置了新的大型窗户,结合屋顶上原有的天窗,提供了充足的日光和河面上壮丽的全景。

"这是一种不寻常的办公楼布局,但是它却有一些显著的优势,使用者没有被引导着远离公司,而是不断和它的空间和社交中心接触。这鼓励了人们的相互往来和参与,也赋予大厅最佳的空间张力;桥梁、地下通道和楼梯意味着人们可以随意流连,这样,他们可以从各个不断变化的角度体验这个空间,并与建筑里面的人们交流,"建筑师Joost Ector这样说,"整个大厅没有空调,只在展馆有,这样能源消耗也减少到最低限度。此外,他们决定使用质量轻、可回收的材料,并将原有建筑作为基础以及促进周边区域的有力因素,以制造一个可持续发展的项目。"

IMd Consulting Engineers

Steel plant becomes "playground for engineers"

The new premises of engineering consultancy firm IMd were opened recently in Rotterdam by alderman Jeanette Baljeu. IMd did not choose a run-of-the-mill working environment: a former

steel plant was transformed in an unorthodox manner into a "playground for engineers", as architect Joost Ector of the Rotterdam firm Ector Hoogstad Architects (EHA) calls it.

Recycling is a big issue in the Netherlands today. A large proportion of the building stock is vacant, awaiting renovation or re-allocation, including premises with unsuspected qualities just waiting for people with initiative who can spot this potential. So does this steel plant on Rotterdam's Piekstraat. It is not an obvious location for an office, but enjoying a unique position with views over the river Maas. What made the building attractive to IMd was the vast space, dominated by an imposing steel structure.

Ector Hoogstad Architects (EHA) and IMd had already worked together on a large number of projects. IMd was also called in when EHA designed an office for itself in a former school building. This last collaboration inspired owners Remko Wiltjer and Pim Peters to look for unique premises for their firm too. They not only saw the

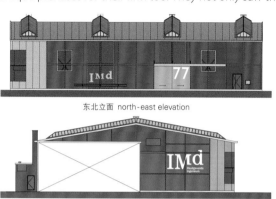

东北立面 north-east elevation

西南立面 south-west elevation

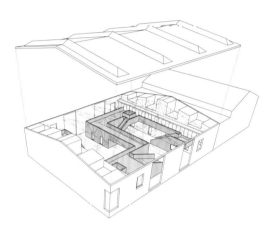

original situation
factory of 1950s with different additions

intervention 1
restore the original robust volume and remove additions

intervention 2
add extra wall openings for daylight and views

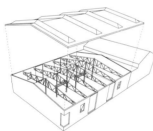
intervention 3
clean and repair characteristic steel structures and facades

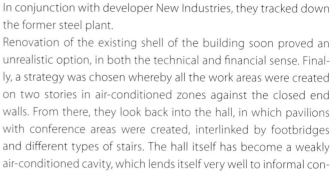

intervention 4
create two office blocks with air-conditioned zones

intervention 5
connect office blocks through bridges and pavilions

advantages of their own organization, but also realized that a striking property would help IMd to position itself even more clearly as one of the leading design engineering firms in the Netherlands. In conjunction with developer New Industries, they tracked down the former steel plant.

Renovation of the existing shell of the building soon proved an unrealistic option, in both the technical and financial sense. Finally, a strategy was chosen whereby all the work areas were created on two stories in air-conditioned zones against the closed end walls. From there, they look back into the hall, in which pavilions with conference areas were created, interlinked by footbridges and different types of stairs. The hall itself has become a weakly air-conditioned cavity, which lends itself very well to informal consultations, lectures, exhibitions and lunching, for instance. Large new windows in what was originally a closed facade, in combination with the existing skylights in the roof, provide daylight and magnificent panoramas over the water.

"It is an unusual layout for an office building, but it does have some big advantages. Users are not directed away from the organization, but are continually in contact with its spatial and social heart. That stimulates encounter and involvement. It also gives the hall an optimum spatial tension: bridges, underpasses and stairs which mean that you can stray and, in this way, experience the space and the people within it from ever-changing perspectives", according to architect Joost Ector. "By not air-conditioning the whole hall, but just the pavilions, energy consumption was also reduced to a minimum. Combined with the decision to use light, recyclable materials, an existing building is as basis and the positive boost for the surrounding area, to produce an extremely sustainable project."

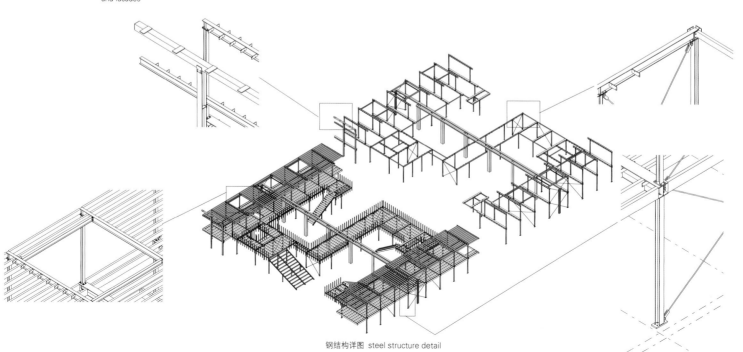
钢结构详图 steel structure detail

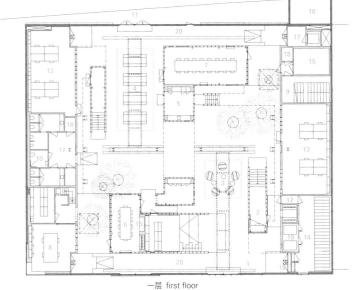

一层 first floor				二层 second floor

1 入口	6 小会议室	11 露台	16 配餐室
2 接待处	7 大会议室	12 人才培养区	17 接待室
3 等候室	8 行政处	13 多功能工作区	18 技术室
4 野餐桌	9 档案室	14 自行车公园	19 储物柜
5 厨房	10 残障人士洗手间	15 服务器	20 中庭

1. entrance	7. large meeting room	12. incubator workspace	16. pantry
2. reception	8. administration	13. multifunctional workspace	17. anteroom
3. waiting room	9. archives	14. bike park	18. technique
4. picnic tables	10. disabled toilet	15. server	19. container
5. kitchen	11. terrace		20. atrium
6. small meeting room			

1 复印室	6 配餐室	11 图书馆
2 会议室	7 聚会场所	12 躺椅区
3 咨询室	8 独立工作站	13 空置区
4 开放式工作区	9 群组工作站	14 桥
5 电梯	10 休息室	15 平台

1. copy	6. pantry	11. library
2. boardroom	7. meeting point	12. chaise longue
3. consulting room	8. individual workstation	13. void
4. open workspace	9. group workstation	14. bridge
5. elevator	10. lounge	15. plateau

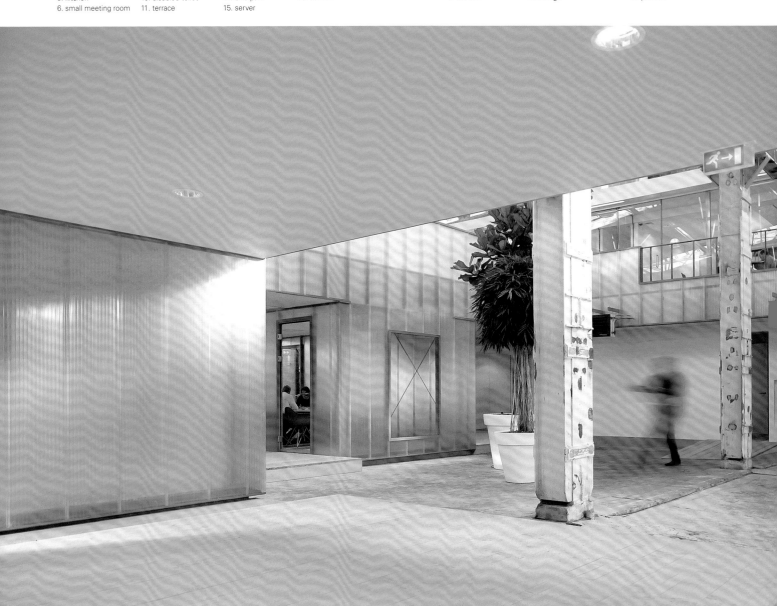

项目名称：IMd Consulting Engineers
地点：77 Piekstraat, Rotterdam
建筑师：Ector Hoogstad Architects
项目团队：Joost Ector, Max Pape, Chris Arts, Markus Clarijs,
Hetty Mommersteeg, Arja Hoogstad, Paul Sanders, Roel Wildervanck,
Ridwan Tehupelasury
结构工程师：IMd Consulting Engineers
安装设计：Unica, Bodegraven
建筑物理咨询：LBP Sight, Nieuwegein
固定装备：Interior architects L.P. van Vliet, Bergschenhoek
(sub-contractor of De Combi)
家具设计：Ector Hoogstad Architects
电气工程师：Unica, Bodegraven
灯光工程师：Muuto, Philips and Lightyears via FormFocus, Zeist
墙壁及门：Qbic and Rodeca, Alphen a/d Rijn
地板：Bolon via Brandt bv, Oosterhout, Ege via Onstein Textiel, Agenturen, Blaricum
可移动设备：Drentea, Feek, Vitra, Wilkhahn and AVL via PVO Interieur zh, Pijnacker
承包商：De Combi, The Hague
总楼面面积：2,014m²
造价：EUR 1,785,000
设计时间：2010.8
施工时间：2011.1—2011.8
摄影师：©Petra Appelhof

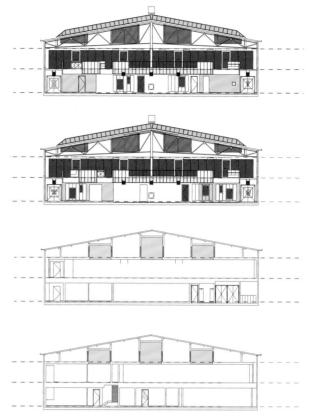

内部墙体示意图1 view of the inner walls 1

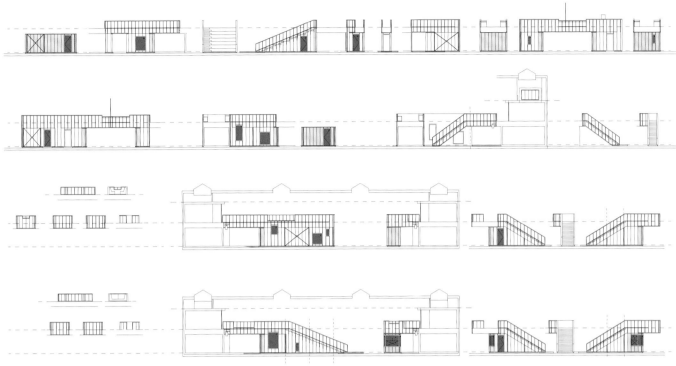
内部墙体示意图2 view of the inner walls 2

红牛公司新总部
Sid Lee Architecture

野兽的诞生

2009年,阿姆斯特丹的红牛公司虽然在荷兰首都周围设有办公室,但仍希望可以搬到更城市化的地方,以期更好地反映它的企业文化和在艺术和体育方面的参与性。

"在内部空间的设计上,我们的目标是重拾红牛理念,根据空间的用途和精神来进行划分;红牛理念有多对相反和相成的方面,像理性与直觉、艺术与工业化城市、黑与白、天使的崛起(红牛给予它翅膀)与野兽的出现……" Jean Pelland (项目设计建筑师和资深合伙人) 说道。在这座包含三处相邻区域的造船工厂里,建筑师着重表达空间上的二分法,从公共空间到私人空间,从黑到白,然后从白到黑。

特征不明确的建筑

这种室内建筑展示了心灵和身体、娱乐和工作、社交和创新式私密性,这种两重性确定了红牛的理念。建筑师通过对立的公共和私人空间来展示这种矛盾。建筑里面三处相邻区域的第一处区域完全是公共空间,其他的两处区域则包括经理办公室和工作站。

斯特拉托斯:异物

在工作区域的中央,矗立着一座庞大的建筑物,它是这座特征不明确的建筑结构的象征。这座穿孔的黑色金属盒状建筑通过它的形状、质地以及褪去的颜色与周围环境结合。这间会议室的灵感来自于屋顶的形状,它的体量就像这座建筑物结构的缩影。

游乐场

这座特征不明确的建筑使建筑师更近一步地游戏于粗犷的体量和不具有垂直性或水平性的折线之间。他们通过多种建筑元素外形的活力和整座建筑原始的、类似地质元素的简朴性,建立起这种语言。因此,这样建立起来的空间对建筑师一贯的视角和标准提出了质疑。为了增强材料的质感,建筑师使用了大量简单的胶合板和未加工的金属板。与这种特意简化的空间布局相反,墙体和地板上布满了非常有趣的图形作品,这些都源自红牛的文化,而后由Sid Lee建筑事务所的阿姆斯特丹团队进行设计。

反叛文化

街头和极限运动文化在红牛理念中得到展现,其特点是有表现欲但是不尊重规则。Sid Lee建筑事务所选择在垂直和水平面上——天花板、墙体、地板——甚至家具上使用独特的图形覆层来表现这一理念。因此,一些房间的墙上遍布涂鸦痕迹。建筑师也希望突出红牛娱乐领域和运动表现,而不是简单地表现这家公司反叛的一面。

多功能空间

作为一处工作区域,同时也是实践艺术和进行运动的地方,位于阿姆斯特丹的红牛公司总部提供了双重功能:工作和娱乐。这里的一切都是多功能的!普通房间里的组合家具可以形成不同类型的座位,但是也可以堆积起来,使这个空间从游乐场变成宴会厅。

智慧和企业精神

建筑师创造这个多功能空间不仅仅是为了展示红牛公司的双重企业文化,他们也想传达源源不断的灵感,大部分得到红牛公司的支持,例如他们知道如何用三个小木头组件建造出BMX坡道,或者如何将一座城市变成"跑酷"的地方。红牛公司的内部空间设计也有同样的精神:一条长凳被移走会形成一个存储空间;墙搁板慢慢拉开时如同一个巨大的抽屉,呈现出一个衣柜。这种可以适应各种情况的能力也显示出其在这个空间内很好地适应下来的天赋。

Red Bull New Headquarters

The birth of the beast

In 2009, with offices located in the surroundings of the Dutch capital, Red Bull Amsterdam wanted to move to a more urban location that would better reflect its culture and involvement in the arts and sports.

"To design the inner space, we aimed at retrieving Red Bull's philosophy, dividing spaces according to their use and spirit, with its two opposed and complimentary hemispheres, reason versus intuition, arts versus the industrial city, black versus white, the rise of the angel (Red Bull gives it wings) versus the mention of the beast…", Jean Pelland, project design architect and senior partner. Inside this shipbuilding factory, with its three adjacent bays, architects focused on expressing the dichotomy of space, shifting from public spaces to

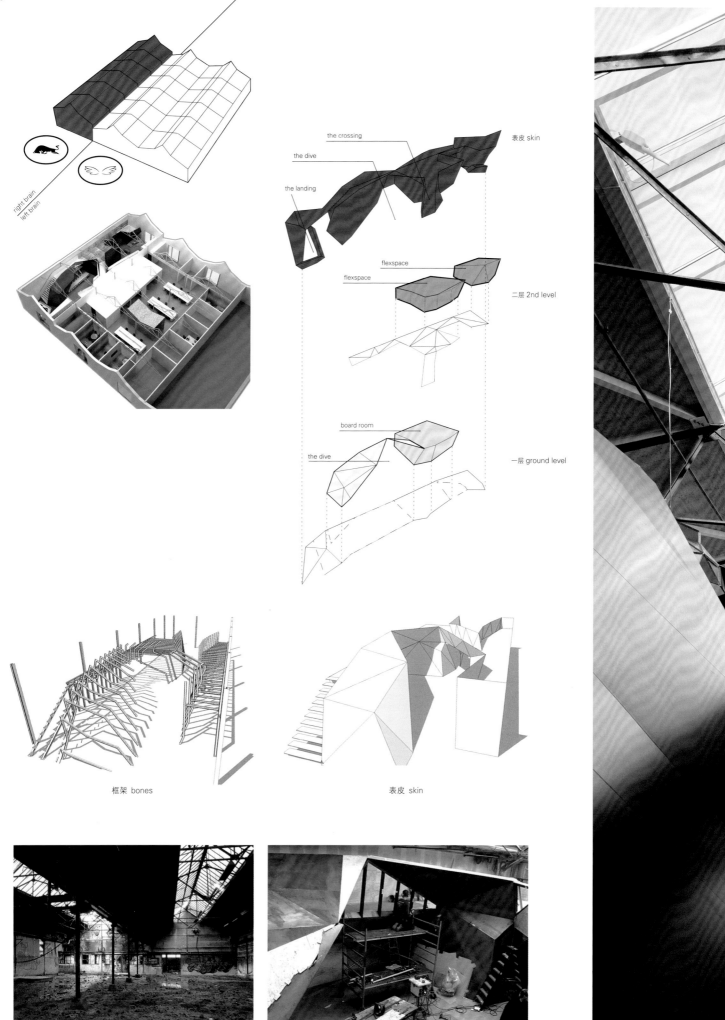

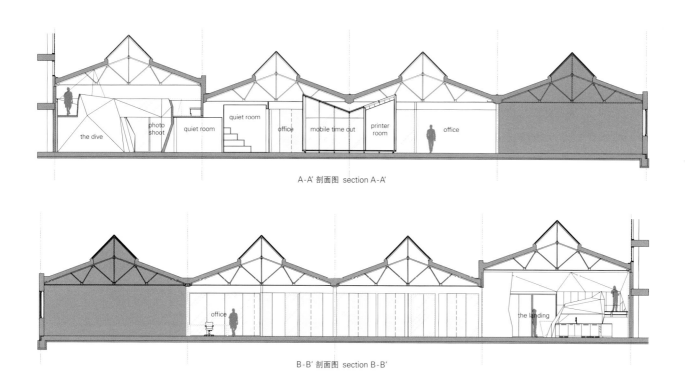

A-A' 剖面图 section A-A'

B-B' 剖面图 section B-B'

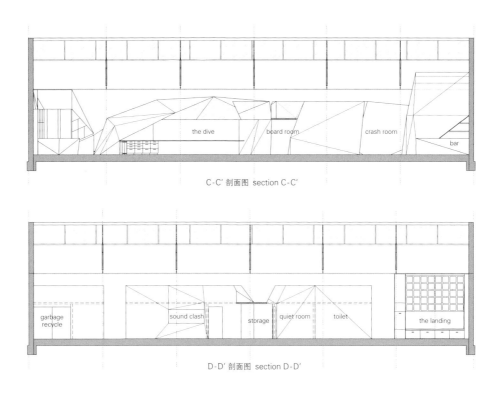

C-C' 剖面图 section C-C'

D-D' 剖面图 section D-D'

private ones, from black to white and from white to black.

The architecture of ambiguity
This interior architecture shows the duality between mind and body, play and work, socialization and creative privacy which defines Red Bull's philosophy. Architects started expressing this ambivalence through an opposition between public and private spaces. The first of the three bays of the building is utterly dedicated to public spaces, whereas the two others contain the managers' offices and workstations.

Stratos: the foreign object
Located in the middle of the working area, a massive architectural object emerges as a symbol of this architecture of ambiguity. This perforated black metal box interacts with the surrounding space through its shape, texture and absence of color. Inspired by the shape of the roof, the volume of this meeting room is like a photo-negative of the building structure itself.

Playground
This architecture of ambiguity led architects even further, playing with brutal volumes and broken lines that refuse verticality and horizontality. They built up this language through a formal dynamism of architectural elements and the primitive, almost geological simplicity of the whole. The space thus created questions of their usual perspectives and marks. In order to strengthen the material brutality, they largely used these elements, such as simple plywood and raw metal plates. In opposition to this intentional simplicity of space layouts, walls and floors were covered with playful graphic works, stemming from Red Bull's culture and further developed by the Sid Lee Amsterdam team.

项目名称：Red Bull New Headquarters
地点：Red Bull Nederland B.V. NDSM-Plein 26, Neveritaweg 34 1033 WB Amsterdam
建筑师：Sid Lee Architecture
当地建筑师（经认证）：Kamstra Architecten BNA
视觉特征和图形：Sid Lee Amsterdam
建筑商：Fiction Factory
家具：2D&W
总承包商：Jora Vision B.V.
甲方：Red Bull Netherlands
用途：Office
总表面积：9,420m² (公共面积：3,058m²，私人面积：6,662m²，夹层面积：870m²)
造价：USD 1,500,000
施工时间：2009.5—2011.4
摄影师：©Ewout Huibers (courtesy of the architect)

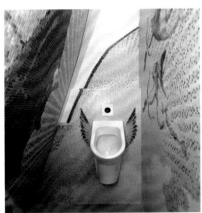

Rebel culture

The street and extreme sports cultures are illustrated in Red Bull's ethic, characterized by a will to perform and not to respect rules. The Sid Lee Architecture team chose to express this philosophy through a ubiquitous graphic covering of vertical and horizontal surfaces: ceilings, walls, floors – and even furniture. Thus, graffiti are all over the walls in some rooms. Architects also wished to highlight Red Bull's universe of play and sport performances, instead of simply expressing the rebel side of the company.

A multipurpose space

As much a working area as a place to experiment arts and sports, Red Bull Amsterdam's headquarters were bound to offer a double function: work and play. Everything is multipurpose here! The modular furniture in the common room forms various seats, but can also be piled up to turn the place from a playground into a party room.

Wits and company spirit

Architects did not create this multipurpose space just to express Red Bull's cultural ambivalence. Their idea was also to convey the idea of resourcefulness, which is so much part of those that Red Bull supports, who know how to build a BMX ramp with three little wooden pieces or change a city into an urban "parkour". Red Bull's interior space was designed in the same spirit: a bench will disappear to give birth to a storing area; a wall shelf will open like a giant drawer to unveil a wardrobe. And this ability to adapt to all kinds of situations also reveals itself in a talent for taking over a place. Sid Lee Architecture

1 地面	5 洒水室	9 服务器机房	13 储存室
2 天井	6 酒吧	10 办公室	14 噪音处理
3 会议室	7 摄影区	11 卫生间	15 非固定休息区
4 收藏室	8 垃圾回收处	12 静室	16 印刷室
1. the landing	5. sprinkler	9. server room	13. storage
2. the dive	6. bar	10. office	14. sound clash
3. board room	7. photo shoot	11. toilet	15. mobile timeout
4. collection room	8. garbage recycle	12. quiet room	16. printer room

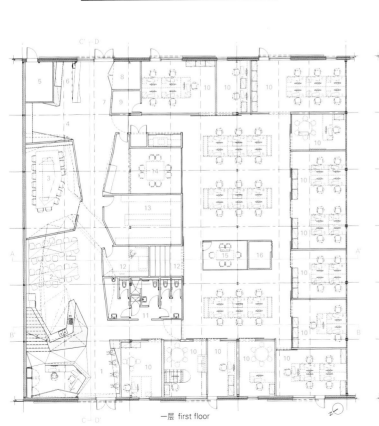

一层 first floor

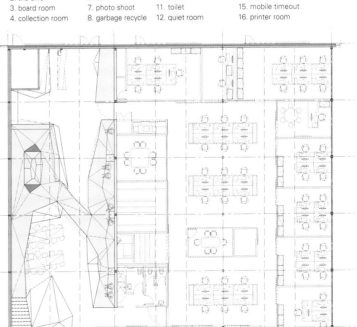

夹层 mezzanine floor

Tesa 105改建项目

Andrés Holguín Torres + David R. Morales Hernández

威尼斯兵船厂是有着非凡历史意义的纪念性建筑综合体。

这座在12世纪发展起来的城市造船厂，经过数世纪的发展成为世界上最大的工厂。

用于生产的建筑和空间一直保持着初始功能，直至第一次世界大战。随着造船体制和方法发生技术变革，它经历了功能性和物理性方面的不断变化。

在20世纪上半叶，由于兵船厂不可能改变空间来适应大规模工业生产的需要，日渐增多的停产区域导致了其结构的普遍退化。

Tesa 105改建项目在2006年威尼斯兵船厂S.P.A.公司发起的国际设计竞赛中胜出，为威尼斯兵船厂北部的复苏、功能再生提供了新的嵌入项目。

Tesa 105构成了兵船厂北部的新入口，同时还因其内部带有相关的服务设施和研究中心办公室而具有入口大厅的功能。

项目的嵌入基于这一准则：预见两种不同建筑风格的交汇，其一是历史的（容器），其二是当代的（内容）。两种结构保持了它们外形的独立，并在相矛盾的元素之间建立起对诘。

原始建筑是16世纪的工业仓库，其外墙将得到完全保留。新体量完全处于建筑内部，它的边界外墙作为建筑的一部分，是一种清晰明确的存在物。

项目的体积方面受兵船厂曾经的造船方法的启发，庞大的船体提升至少数几个支撑系统上，使人们可以在船体底部工作。

类似地，四个小型空间巧妙地分布在一层，向公共入口开放（信息台、书店、会议室和咖啡馆），以支撑一个更大的空间——一家私企的五个办公空间，威尼斯S.P.A.兵船厂办公室都设置在那里。在顶层，两个玻璃体量充当了楼下办公室的会议室。

新结构的空间被转移到古老建筑的一侧，以便屋顶天窗能获得更多的自然光，也为原始建筑布局提供了一处明亮的阅读场所。

建造过程中仅使用了少数几种建材（玻璃、钢材、陶质覆层），材料表面的彩色被减至白色、渐变的灰色，旨在避免与原始建筑裸露的富有表现力的砌砖表面产生视觉冲突。

Conversion Tesa 105

The Venice Arsenale is a monumental complex of exceptional historic importance.

The city's shipyard, having grown during the XII century, developed to become the largest factory in the world for centuries.

The buildings and spaces of production maintained their original function until the beginning of WWI, being subject to the constant functional and physical adaptations that followed the technical evolutions of shipbuilding systems and methods.

During the first half of the 20th century, the impossibility of adapting the spaces of the Arsenale to the needs of large-scale industry brought about a progressive abandon of the area resulting in a generl deterioration of its structures.

The conversion project of the Tesa 105 won the International project design competition launched in 2006 by the Arsenale di Venezia S.P.A. enterprise for the realization of new project interventions for the recovery and refunctionalization of the northern section of the Venice Arsenale.

The Tesa 105 constitutes the new northern access to the Arsenale and will host the entrance hall with its related services and offices for the research centers therein.

The project intervention is based upon criteria that foresee the joining of two different styles of architecture, one that is historic (container) and the other contemporary (content). Both structures preserve their formal autonomy and establish a dialogue among contrasting elements.

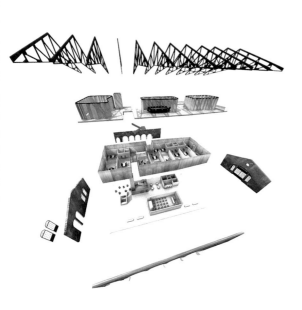

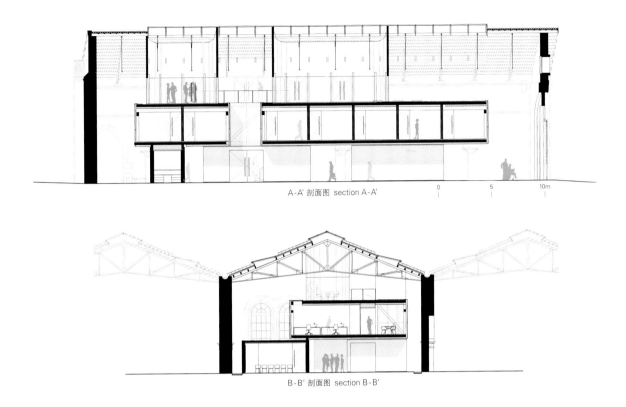

A-A' 剖面图 section A-A'

B-B' 剖面图 section B-B'

1797年威尼斯兵船厂的透视图
Perspective view of the Arsenale of Venice refers to 1797

西望威尼斯兵船厂
West view of the Venice Arsenale

威尼斯和海军1889—1920年
Venice and the Navy 1889~1920

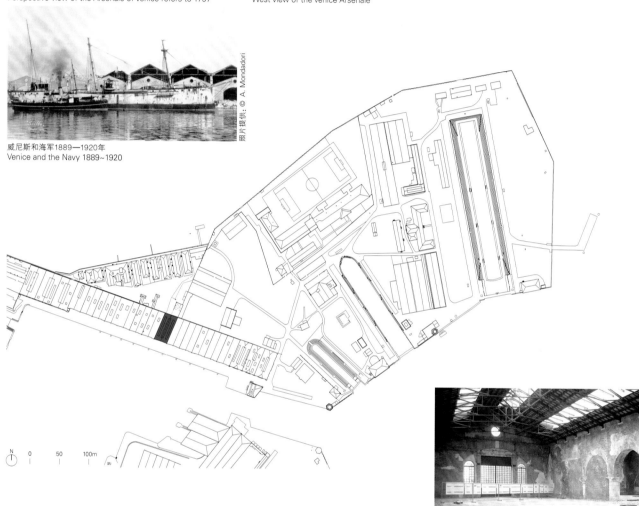

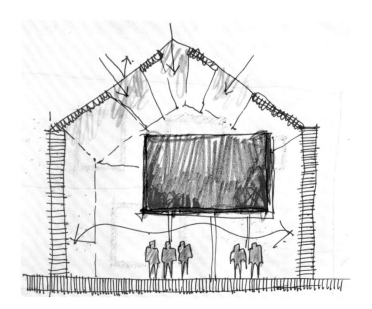

项目名称：Conversion Tesa 105
地点：Arsenale, Venice, Italy
建筑师：Andrés Holguin Torres, David Ricardo Morales Hernández
结构工程师：Thetis Spa, ing. Giovanni Zarotti
甲方：Magistrato alle Acque-Arsenale di Venezia s.p.a.
用地面积：682.75m²
建筑面积：1,130m²
总楼面面积：359.15m²
设计时间：2007—2009
施工时间：2010—2012
造价：EUR 2,800,000
摄影师：
courtesy of the architect - p.104bottom, p.106top-left, p.106middle
©Andrea Pertoldeo (courtesy of the architect) - p.99, p.101,
p.102~103, p.104top, p.105, p.106top-right, p.106bottom

The outer walls of the original building, a 16th century industrial warehouse, will be entirely maintained, while the new volume remains completely contained within the interior of the building, its perimeter walls, as an explicit presence and part of the building.

The volumetric aspect of the project is inspired by the way that the ships were once built in the Arsenale, where the vessels' volumes were elevated onto few support systems allowing for work beneath the ships' hull.

Analogously, four small-scale volumes distributed strategically on the ground floor and open for public access (info point, bookshop, conference room and cafe), are to support a larger scale volume where five office spaces for a private enterprise and the offices of the Società Arsenale di Venezia S.P.A. will be located. On the top floor, two glass volumes will host the main meeting room for the offices below.

The volume of the new structure was moved to one side of the historic building as to allow for more natural light from the roof's skylights and to allow for a clear reading place of the original building's layout.

There are few materials used in the construction (glass, steel, ceramic coatings) and the chromatic range of their surfaces has been reduced (white and gradations of gray), as to avoid creating visual conflict with the expressive surfaces of the original building's exposed brick. Andrés Holguín Torres + David R. Morales Hernández

1 信息台　2 书店　3 接待处　4 寄存间　5 酒吧寄存间
6 浴室　7 酒吧　8 多功能室
1. info point 2. bookshop 3. reception 4. deposit
5. bar deposit 6. bath room 7. bar
8. multipurpose room

一层 first floor

1 浴室　2 服务器机房　3 办公室　4 会议室
1. bath room 2. server room 3. office
4. meeting room

二层 second floor

1 会议室　2 设备间
1. meeting room 2. installation room

三层 thrid floor

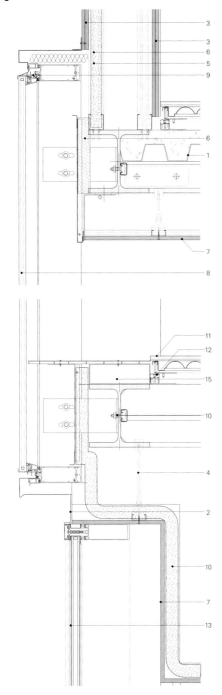

1. fretted metal sheet 2. "L" profile in painted steel 100×30mm 3. water-resistant plasterboard-12.5mm thick 4. aluminium profile structure to fasten the suspended ceiling 5. aluminium profile structure to fasten plasterboards 6. mineral wool thermal insulation-40mm thick 7. suspended ceiling in plasterboard, stuccoed, smoothed and whitewashed 8. double glazing windows composed of exterior glass pane 6+6mm with interposed serigraph film layer 9. extruded profile in coated aluminium alloy 10. fixing plate for fastening casing and structure 11. linoleum covering-2.3mm thick-over the floating flooring elements 12. mechanical spacer for the floated flooring elements 13. double glazing windows composed of external anti-shatter glass pane 6+6mm, 12mm gap and 5mm internal anti-shatter glass pane 14. white epoxy resin 15. aluminium tubing bolted to the structure 16. concrete screed 15cm thick 17. layers of thermal insulation 30mm 18. concrete screed 15cm thick

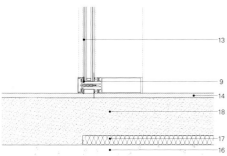

楼梯井详图 stairwell detail

1. whitewashed gravel - 5/10mm in diameter
2. fretted metal sheet
3. load bearing structure in painted steel
4. "L" profile in painted steel 100×30mm
5. water-resistant plasterboard – 12.5mm thick
6. coating in sheets of pliable ceramic covering – 3mm thick
7. filler in lightened concrete
8. aluminium profile structure to fasten the suspended ceiling
9. aluminium profile structure to fasten plasterboards
10. mineral wool thermal insulation - 40mm. thick
11. suspended ceiling in plasterboard, stuccoed, smoothed and whitewashed
12. pavement in wooden cubes
13. double glazing windows composed of exterior glass pane 6+6mm with interposed serigraph film layer
14. extruded profile in coated aluminium alloy
15. fixing plate for fastening casing and structure
16. extruded aluminium profile
17. floating flooring supports
18. steel sheet 60×60
19. layered glass composed of 6mm tempered glass + pvb interlayer + 6mm tempered glass
20. insulating window wall 4+4-12-4mm and relative motorized accessories
21. air brick wall
22. steel tubing Ø=80mm t=4mm, fastened at the base of a concrete slab with chemical anchor bolts
23. linoleum covering - 2.3mm thick - over the floating flooring elements
24. mechanical spacer for the floated flooring elements
25. aluminium profile with fastenings to the casing structure
26. white epoxy resin
27. aluminium tubing bolted to the structure
28. concrete screed 15cm thick
29. layers of thermal insulation 30mm
30. stabilized granular foundation minimum thickness 10cm
31. double water-proof sheathing with slate top-face

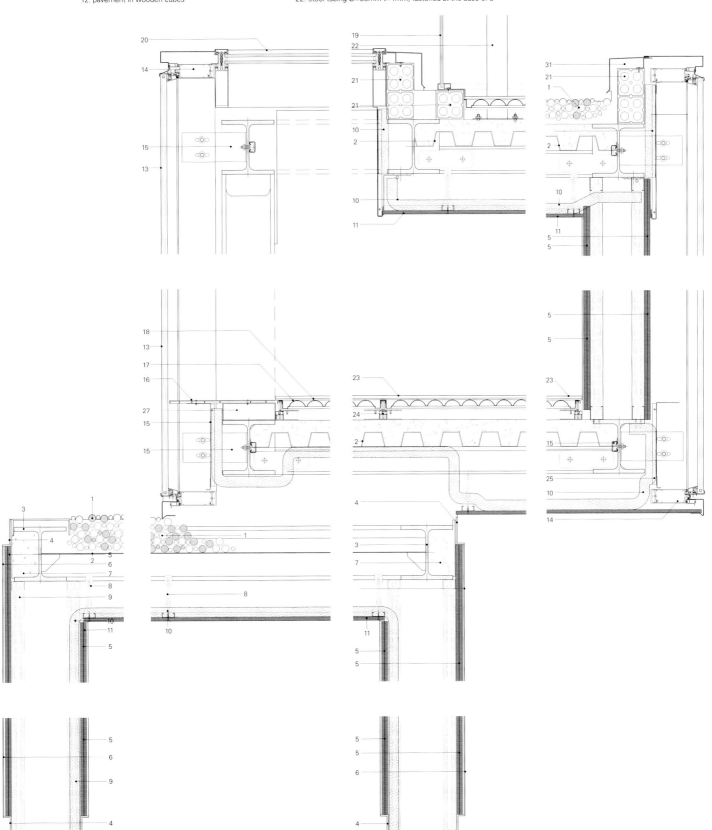

剖面详图 section detail

在景观中创建一种信号

该项目位于圣皮埃尔的一条运河沿岸，即前加来工业区的所在地。它位于运河下游100m处，并延续了由La Cité de la Dentelle (由Moatti & Rivière建筑师事务所建造) 开始的城区改造活动。在这片荒凉的城市景观中，对原有的工业大厅进行的改建必须显而易见。未来的设施必须表明它的存在，并吸引潜在用户、年轻人和猎奇者进入这里。楼前较高的空地使人们从周围区域可以更清晰地望见西面山墙。这面山墙已进行了全面的重新设计，将向公共空间传递出强烈的信号。

Zap' Ados

Bang Architectes

重新划分大厅

原有建筑是一个常见的、没有什么特殊之处的工业大厅，是预制混凝土板和水泥板屋顶填充而成的混凝土结构。大厅曾是一家生产烤花生的工厂，后来具有其他各种用途（包括卡丁车赛道），最终在几年前被遗弃。在接手和处理这座建筑之前，它已经残破不堪，遭到损毁，结构也不稳定。在清理和恢复结构前的第一个任务就是打开黑暗的大厅。建筑师通过移除东立面和西立面的预制混凝土板来开放视野，把自然光带入建筑的核心，以此来完成这一任务。

展现建筑的新任务

青年中心和滑板公园越过了山墙，形成两个凸起结构，这两个结构明显地标志出该建筑有一个新的目标。其中一个凸起结构位于地面，并从滑板俱乐部和青年中心处延伸出来，形成了室内外空间之间的接触点。另一个凸起结构是悬臂式的发射台，俯瞰正面广场，描绘出滑板者出发前等候的形象。两个棱柱形体量如同伸开的手臂，重新对正面广场的自由空间进行划分，做出邀请进入的姿态。其建筑表达通过一个多孔金属网构成的普通围护结构统一起来，这个围护结构将建筑轮廓由挂钩形变成一个从单一混合形状中突出的棱柱形。这张金属网允许观众观看里面的活动，并且从上到下逐渐穿孔。金属网如同一扇百叶，色彩鲜艳，且控制着直射阳光的进入量。这是有意突出的设计因素。这张多彩的金属网给设备提供了保护，因为这种多孔金属网十分坚固，且具有良好的抗涂鸦性。网格为双层，并带有幕墙，保护用户免受盛行风的干扰，降低了滋扰附近居民的噪音。楼外的正面广场用橙色的框标出停车空间，与现有地面涂层的纹理相重叠。

连接两个项目

大厅内的各种项目元素都进行了纵向规划，以使滑板轨道的长度最优化，并且沿南立面为青年中心设置了一个入口。进入大楼时，人们能看见一系列设置在右边的封闭供暖室，它们可通过外露的棱柱形状辨认出来。这个设施带有轻质框架，被放置在原有的平板上，入口沿着室内通道分布。该通道完全融入滑板公园的空间中，并用一个和大厅全长等长的栏杆分隔开来。这条通道能够使旁观者安全地观看溜冰者。长长的墙壁上覆盖着吸音织物，织物延展开来，形成巨大的"涟漪"，这种吸收表面的设计可减少由滑板在坚硬地面上发出的声音回响。墙体还为滑板公园和青少年中心的用户提供用于观看的垂直窗口。大型声音挡板的正弦层从天花板悬垂下来，为用户增加音质舒适感，这些技术性构件提供了经济实惠的改进，改变了内部的空间，并隐藏了不雅观的天花板。

创建滑板轨道

模块呈带状，沿大厅全长设置。西侧一个凸起的平台可以俯瞰正面的广场。平台可以作为一个高点：发射台。碗状装置（在该地区罕见）设置在大厅的东端，以保持空间净空。这些复杂曲面是木工和木匠精工细作的完美作品。大厅中央为娱乐场地。一处安静的试滑区沿着室内通道设置，其间穿插着模块。模块是由木材（非混凝土）制成，以保持滑板公园的适应性和大厅原有配置的可逆性。

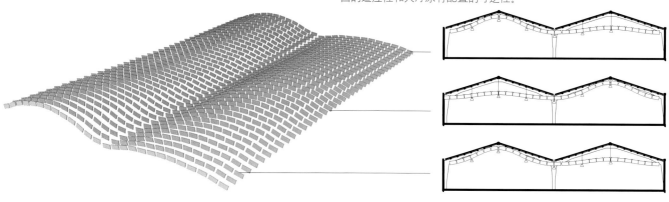

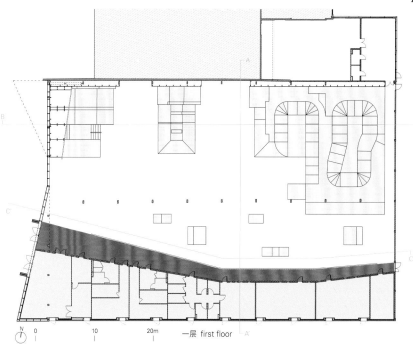

一层 first floor

西立面 west elevation

Zap' Ados

Create a signal in the landscape
The operation takes place along a canal in St. Pierre, which is the former industrial district of Calais. It continues the urban renewal initiated by La Cité de la Dentelle (by Moatti & Rivière Architects) and is located a hundred meters downstream. In this bleak urban landscape, the conversion of the existing industrial hall has to be visible. The future facility must signal its presence and invite potential users, the young and curious, to enter. The high clearance at the front of the building offers increased visibility of the west gable from the surrounding area. This gable, which has been completely redesigned, will project a strong signal into the public space.

Reclassify the hall
The existing building is a common industrial hall with no outstanding features, consisting of a concrete structure filled with precast concrete panels and a roof of cement sheets. The hall was once a roasted peanut factory, followed by various other incarnations (including a go-kart track) before being abandoned for several years. Prior to handing and processing it had been dilapidated, vandalised and had become structurally unsafe. The first task was to open the dark hall before curettage and structural recovery. This was achieved by removing precast concrete panels on the eastern and western facades to release through-views and bring natural light into the heart of the building.

Express the new assignment of the building
The youth centre and the skate park extend beyond the gable and form two protrusions, which clearly signify that the building has a new purpose. One protrusion stands on the floor and emerges from the skateboarder club and youth centre, forming a point of contact between the inside and outside space. The other is cantilevered launch pad that overlooks the front square, featuring skaters waiting in turn before taking off. The two prismatic volumes, like opened arms, reclassify the free space of the front square and act as an invitation to enter. The architectural expression is unified by a common envelope made of expanded metal, which turns the silhouette from a hanger into a prism protruding from a singular hybrid form. The metal mesh allows spectators to watch activities inside and is gradually perforated from top to bottom. The mesh acts like a shutter, controlling direct sunlight and the color is stricking. It is deliberately conspicuous. This colorful mesh protects the equipment as the expanded metal is very resistant and anti-graffiti. It is doubled with a curtain wall to protect users from prevailing winds and reduce any noise nuisance to nearby houses. Outside the building the front square is treated using an orange frame to draw parking spaces, which overlap the textures of the existing floor coatings.

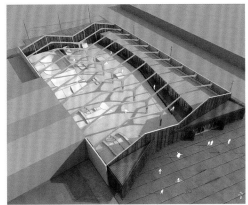

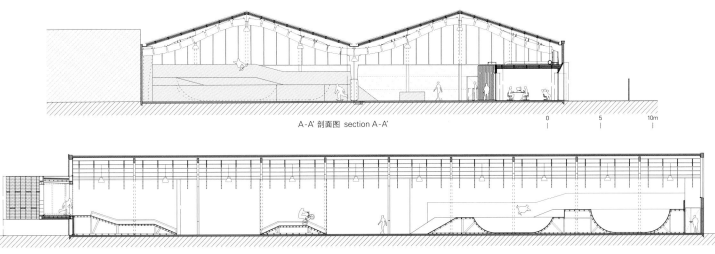

A-A' 剖面图 section A-A'

B-B' 剖面图 section B-B'

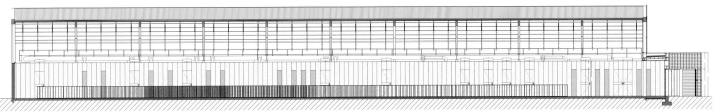

C-C' 剖面图 section C-C'

Linking the two programs

Inside the hall the various program elements are organised longitudinally, to optimise the length of the skate tracks and provide an entrance to the youth centre along the southern facade. When entering the building, there are a series of enclosed and heated rooms installed on the right identified by emerging prism. This set is built with a light frame and placed on the existing slab, with entrances distributed along an indoor walkway. This walkway is fully integrated into the space used for the skate park, separated by a handrail that runs its entire length. It enables "spectators" to watch the skaters safely. The long wall is covered with an acoustic fabric stretched to form large "dimples". This absorbing surface is designed to reduce reverberated sounds caused by skateboarding on hard surfaces. The wall is also provided with vertical windows offering views for both users of the skate park and youth centre. A sinusoid layer of large acoustic baffles is suspended from the ceiling to increase acoustic comfort for users. These technical elements offer inexpensive modifications that morph the inner space and hide the unsightly ceiling.

Create skate tracks

The modules are arranged in a strip oriented along the full length of the hall. On the west side a raised platform overlooks the front square. It serves as a high point: the launcher. The bowls (rare in the region) are installed at the east end of the hall to maintain space clearance. These complex curved surfaces are works of joinery and carpentry of great sophistication. In the centre of the hall is the funbox. A calm initiation zone is arranged along the indoor walkway and punctuated by modules. The modules are made of wood (not concrete) to maintain the adaptability of the skate park and the reversibility of the original allocation of the hall.

Bang Architectes

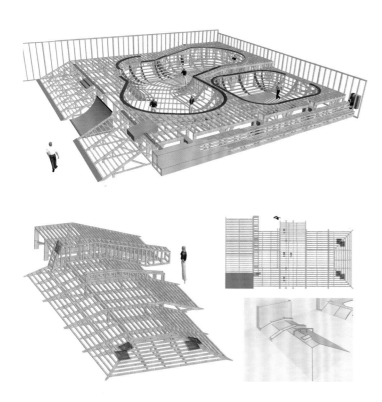

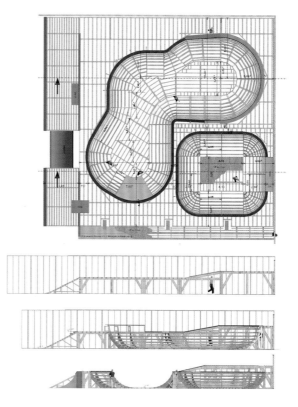

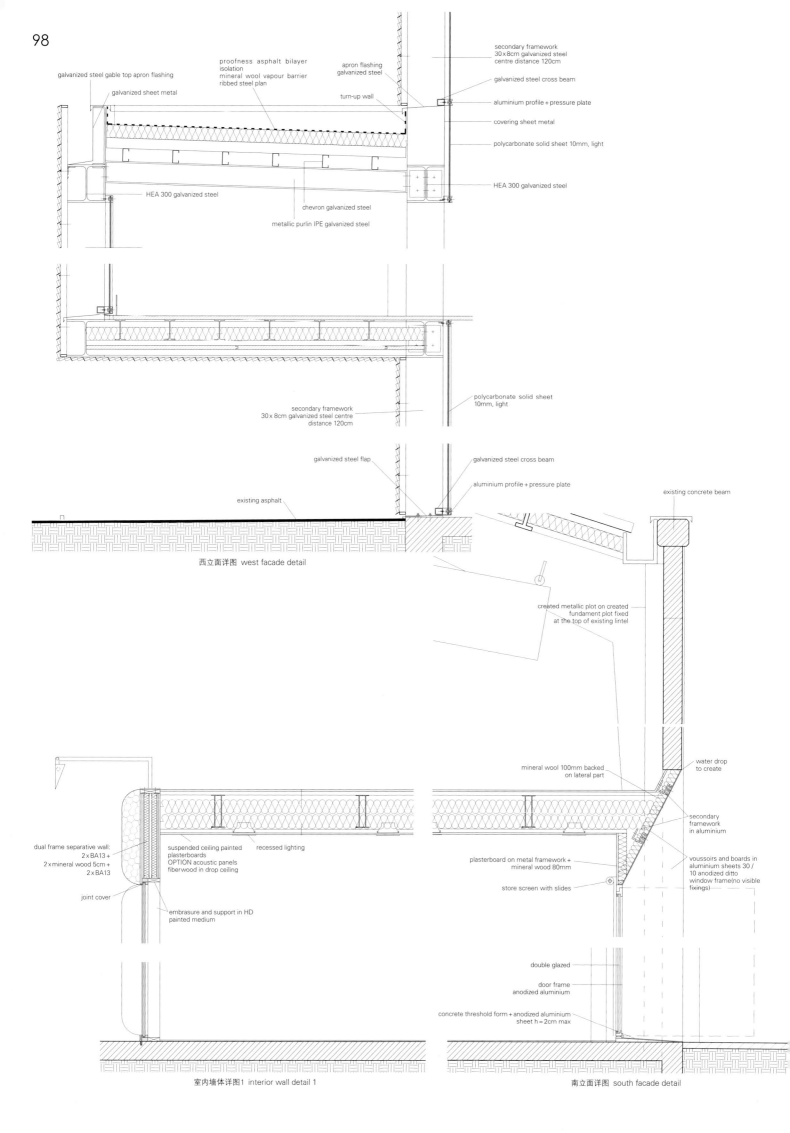

项目名称：Zap' Ados
地点：87 quai de Lucien L'heureux 62100 Calais
建筑师：Nicolas Gaudard, Nicolas Hugoo
工程师：B&R ingénierie
甲方：Ville de Calais
用途：skate park, youth centre
总楼面面积：2,760m²
设计时间：2010.6
竣工时间：2011.12
造价：EUR 1,500,000 H.T.
摄影师：©Julien Lanoo (courtesy of the architect)

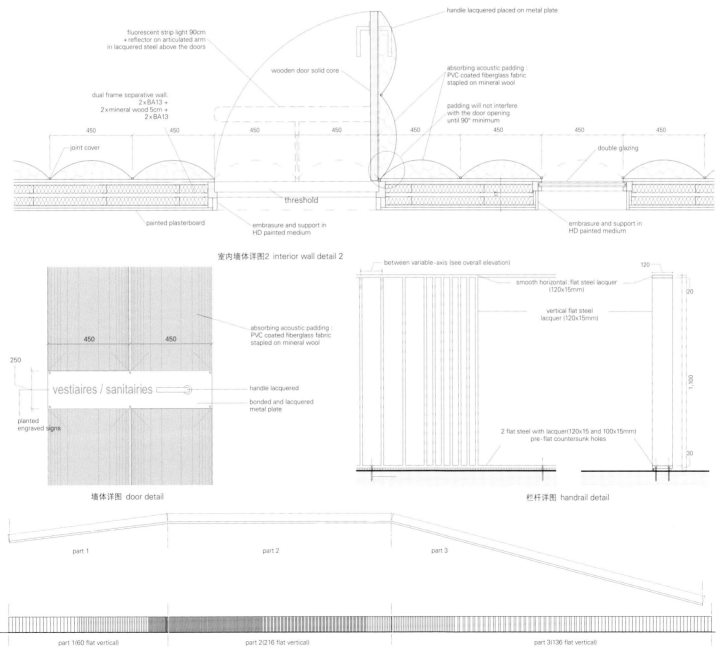

室内墙体详图2 interior wall detail 2

墙体详图 door detail

栏杆详图 handrail detail

栏杆立面 handrail elevation

米诺——意大利米利亚里诺青年旅社

Antonio Ravalli Architetti

在将老旧的麻厂转变为米利亚里诺镇新城市中心区的进程中，青年旅社项目争取到整个建筑面积的510m²。项目由于临近波河三角洲自然公园，所以处于夏季环游线路的重心，但该项目还是必须依靠减少了的当地资金（270 000欧元）来开展，这笔资金已经囊括了家具设施费用以及待定的管理盈利。因此，项目管理方面（能源和资金节约方面）显得尤为重要。该青年旅社被构想为一座"消极机器"，室内连续不断输入新鲜空气，以获得良好的气候条件；而在客房的布局与形态的设计方面，经过构想，使用了最少的构件和技术，从而具备了弹性容纳能力：春夏两季或者举办特殊活动时达到最高，而在淡季只能够维持基本保障。

一层包含了接待处和各类设施，二层则是统一单侧开窗的独立巨型空间。4间客房（每间可容纳2~3人）、浴室与楼梯分布在两层与窗户相对的一侧。这些空间构成了一种紧凑而轮廓分明的体量，可以用传统设备进行空气调节。而在主空间内，空气调节基于被动式通风设备，即通过北侧的窗户和屋顶上的两座通风塔来进行。由于光与空气传播途径的特殊性，空间不可能被分割成诸多单元，这给青年旅社的集体宿舍提供了一种备选方案：建筑师在这里设置了类似于室内露营区的独立小单人间，单人间外包裹着轻质材料。这些房间不仅形体独立，室内气候上也独立：一种定时的空调系统可以选择面向哪几间"房间"打开。整个系统网处在可检视的木质平台的下部，将各个小单人间联系起来。空间高度的变化标志着"房间"中较为私密的空间向日常公共活动区的过渡。

平台周边的建筑活动创造出就座和放松的舒适环境，舒适的客房适合人们在此阅读或上网。尽管容纳了多种不同的用途，空间依然显得很流畅，而嵌入的构件和家具的单色性更增强了它的可塑性。

MiNO, Migliarino Youth Hostel

As part of a program for the conversion of an old hemp factory into a new city center for the town of Migliarino, the project gains a youth hostel out of a 510m² portion of the building. The site position is barycentric to the touristic circuits which take place during the summer, thanks to the proximity of the Po River Delta Natural Park, but the project has to count on a reduced regional funding, 270,000 €, including the furniture, and a doubtful management profitability. Thus the management aspects, both with the energetic and economic saving, are the principal matters. The hostel is imagined as a "passive machine", in which natural air fluxes are conveyed in order to obtain climatic benefits, while the systems distribution and the morphological disposition of the rooms, conceived as to minimize the utilized elements and technologies, allow an elastic hosting capacity: the highest during the spring and the summer, or in case of special events, and reduced to the essential during the low seasons.

While the reception and the facilities are located on the first floor, the second level presents a single big space with all the windows on just one side. Four rooms for 2-3 persons each, bathrooms and a staircase are disposed on two levels on the other side. These compose a volume which, compact and well-defined, can be air-conditioned with traditional tools. In the main space instead, the air conditioning is based on passive ventilation, eased by the po-

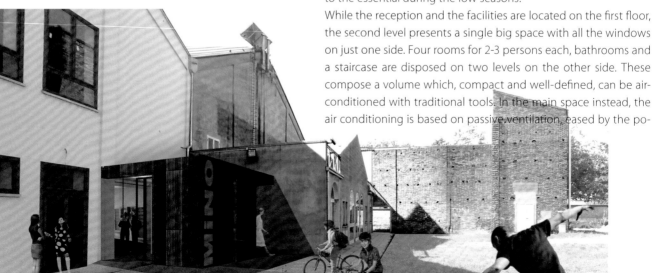

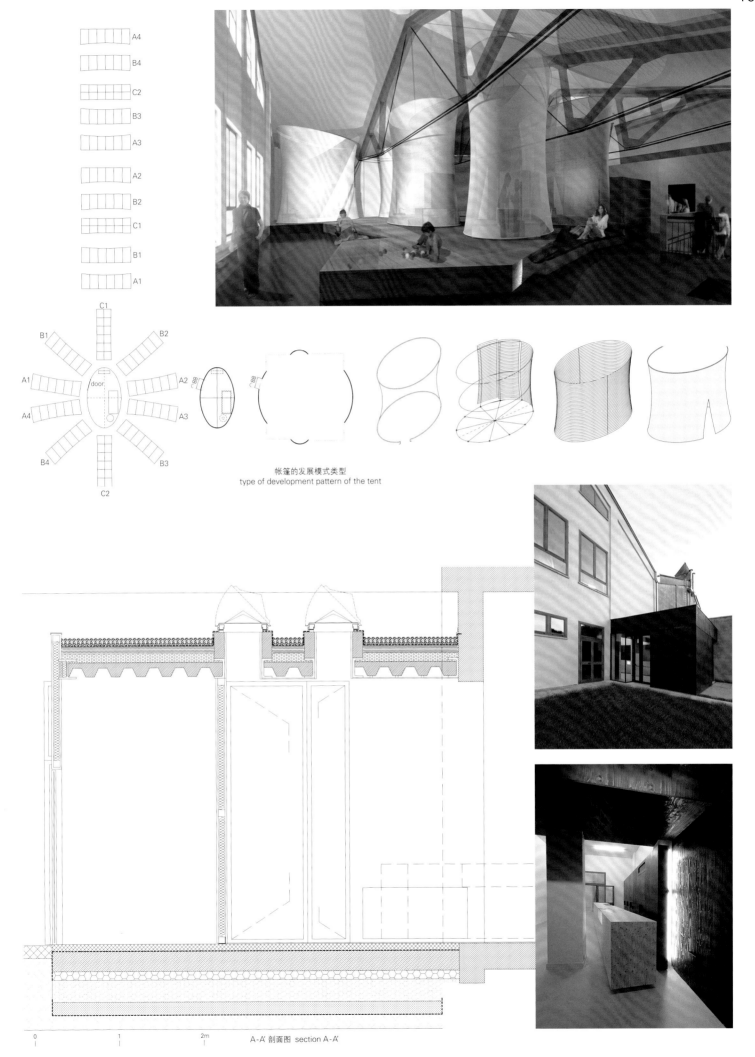

帐篷的发展模式类型
type of development pattern of the tent

A-A' 剖面图 section A-A'

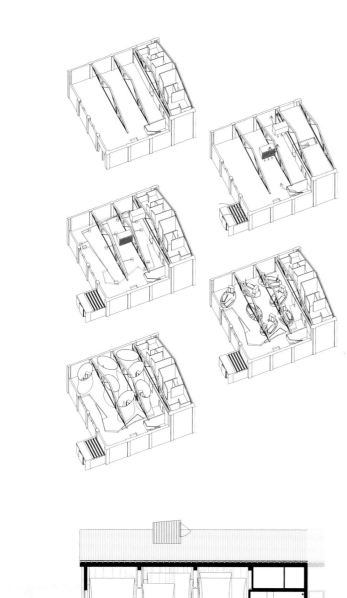

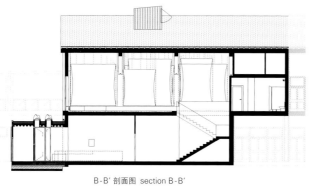

B-B' 剖面图 section B-B'

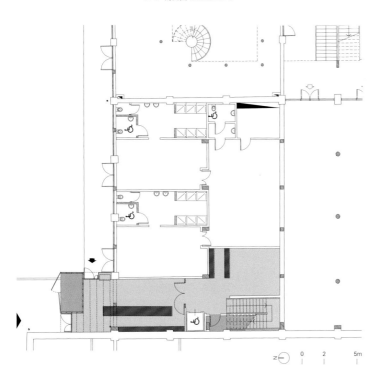

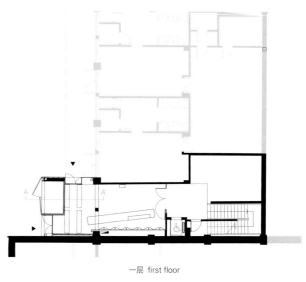

一层 first floor

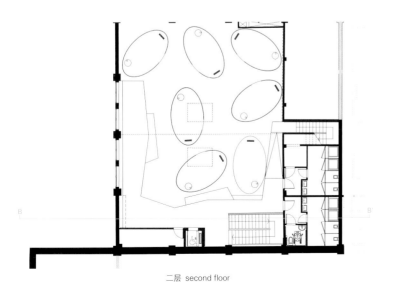

二层 second floor

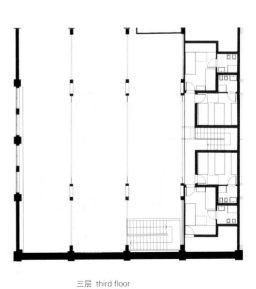

三层 third floor

sition of the windows on the north side and by two ventilation tower located on the roof. The impossibility of this space to be divided in more units, due to the uniqueness of light and air provenance, suggests an alternative solution to the dormitory: like an indoor camping, autonomous cells are placed, enfolded in light wrapping. They are not just physically, but also climatically independent "rooms": a punctual air-conditioning system permits to choose which ones to "turn on". The entire system net is located under the inspectable wood platform, which works as a connective tissue for the cells. The difference in height marks the transition from the more intimate space of the "rooms" to the common daily area. The movement of the platform perimeter creates occasions for sitting and relaxing, cozy niches in which to read or surf the net. The space remains fluid, though allowing a multiplicity of distinct uses, while its plasticity is enhanced by the monochromaticity of the introduced elements and furnitures. Antonio Ravalli Architetti

项目名称：MiNO, Migliarino Hostel
地点：Migliarino, Ferrara, Italy
建筑师：Antonio Ravalli, Simone Pelliconi, Valentina Milani, Lorenzo Masini, Giuseppe Crispino
结构工程师：Mezzadringegneria srl
电气工程师：GF studio associato
空调系统：Studio Zambonini　承包商：Aurora costruzioni S.A.S.
甲方：Comune di Migliarino (FE)　用途：Hostel　建筑面积：510m²
材料：wood, concrete, fabric　竣工时间：2010

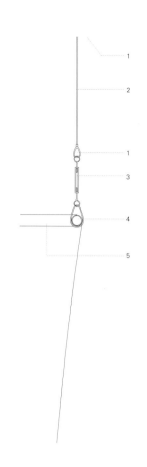

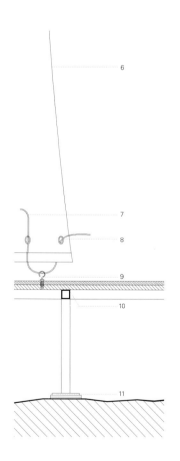

1. splice
2. spiral steel cable
3. turnbuckle
4. spiral steel cable
5. steel pipe Ø6 cm, length 1,265cm
6. pod: coated polyester fabric 300~250 gr.
7. elastic rope
8. loop
9. eyebolt
10. larch boards (12mm)
11. steel floor structure

旧厂房的空间蜕变 Uptodate in outdated factory

La Fabrique——创新机器
TETRARC

丰富创意的沉淀和多样性才能

La Fabrique位于"南特市的île",在原有的Dubigeon仓库一旁,临近阿尔斯通仓库(即将成为"Quartier de la Création"的中心);老船厂将被改造成大花园。该环境为收集这一被历史遗忘的地方的过去提供了全面的机会,以此提醒我们,文化是由沉积和大量的重新发现组成的连续统一体。通过集合公众及各地表演或室内音乐会的艺术家,重大的项目和倡议由此产生:两座大楼表明了这个独特项目的双重追求,同时TETRARC建筑师事务所建造的建筑彰显了南特市的身份。古老的场地上坐落着一些工业仓库,建造这些仓库采用的技术使科学与美学相融合,创造出一种新的美。船舶的历史是凭借那些拥有熟练技能和才华的工人们的力量建立起来的,艺术家的历史使他们可以用眼睛去认知宇宙——从画全新钢结构的画家莫奈,到习惯于忽略金属形状的贝歇尔。最后,工业生产组织的历史带来了被置于同一个屋顶下的试验车间、装配线、整改车间和出口站,创造了一幅具备灵活性理念的生动画面。该项目包括三个不同部分:容纳400人的大厅(内部设有办公室)、容纳1200人的大厅——两者由一个公共大厅相连,该大厅置于Dubigeon仓库的混凝土柱框架中,最后是悬置于旧防空洞上方的工作室。

建筑的三个时期

Léon Bureau大道引人注目的空间和la Prairie aux Ducs大道交叉口的信号灯,都用来创造该项目(呈现给建筑师)能够被立即感知的三个主要部分:接待区、两间陈列室、办公室和工作室套间。然而,赋予设备形式,可以让公众发现声音和图像(意在设计一种富于表现力和原创艺术的建筑)的探索者创造的新世界。一些经过折叠、组装、堆放、任意组合从而覆盖功能性的室内空间的金属板对上述形式进行了回应。

仓库的连接作用

La Fabrique与Dubigeon仓库毗邻,它利用Dubigeon仓库的混凝土柱形成一个安全的公共空间,里面设有咨询台、自动取款机、产品销售柜台以及酒吧/餐厅。这个长长的、形状不规则的大厅布置得既简洁又明亮,而且与举办"机器和大象的岛屿"活动的大厅形成了一种流畅的连续性。大厅还提供了通往服务该区域的最大购物中心的便捷通道。不久,国立建筑学院和阿尔斯通仓库将被改建成美术学院。

Maxi-Micro厅

大型金属覆盖的全景窗户朝向城市海洋,创造了Léon Bureau大道的部分正面景色,建立了一个召集市民的信标。该建筑使用斜面空间将自身延展,通过锥形长金属板进行调整,并通过半拖车大小的大门来凸显。通过点缀其间的孔眼,人们可以偶尔看到管道和设备,这些孔眼显示了大宗设备的复杂性,这些设备能使1200人在Maxi厅欣赏著名艺术家的表演,并使200位有洞察力的人在Micro厅发现新兴的人才。如果需要的话,这些演出可以由生成多媒体的声波实验平台来提供。

人才的发动机

从一个中空且水平的防空洞的混凝土基础来看,这个起重设备似乎向天空升起,面向未来的带有七层闪光地板的办公楼和带有城市职能的录音棚/排练室。通过动态地嵌入在古老的墙壁和煤仓的材料之间,La Place提供了小团体之间的分享时刻:在彩灯下,文化咖啡厅可以容纳100人,举办小型音乐会和会议。在这上面,La Terrace用其嵌入到建筑中的MAO (Musique Assistée par Ordinateur) 巴士来延伸接待处。

La Fabrique, Creative Machine

Fertilize the presence of creative sedimentations and multiple talents

La Fabrique is located on the "île de Nantes," to one side of the

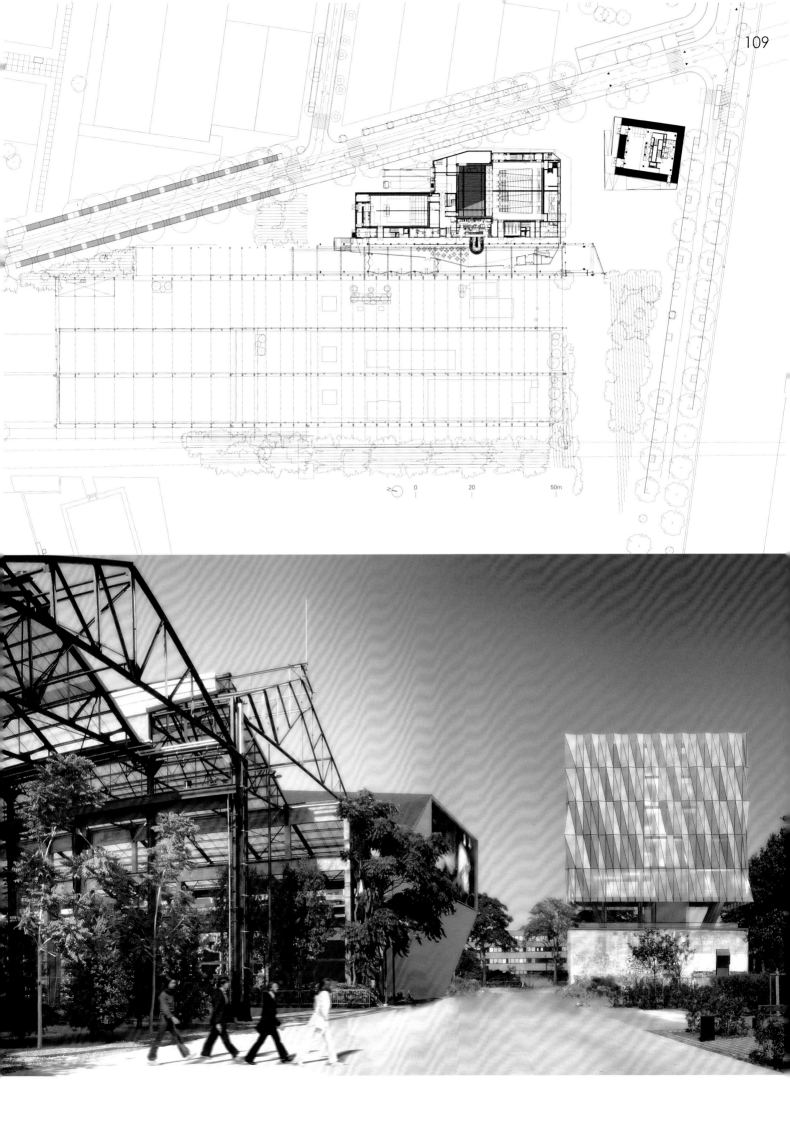

1 工作室 1. studio
五层——A fifth floor_A

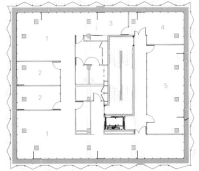

1 办公室 2 个人办公室 3 单人间 4 五人间 5 会议室
1. office 2. personal office 3. personal space
4. space for 5 people 5. meeting room
四层——A fourth floor_A

1 露台 1. terrace
三层——A third floor_A

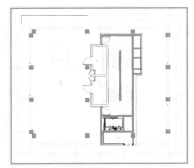

1 中空空间 2 夹层露台
1. void 2. mezzanine terrace
二层——A second floor_A

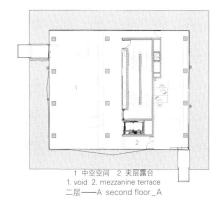

1 信息台 2 夹层露台 3 原有建筑
1. info point 2. mezzanine terrace 3. existing building
一层——A first floor_A

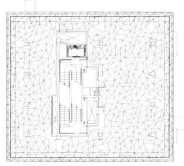

露台平面图R+2 terrace floor plan R+2

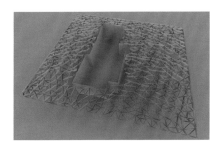

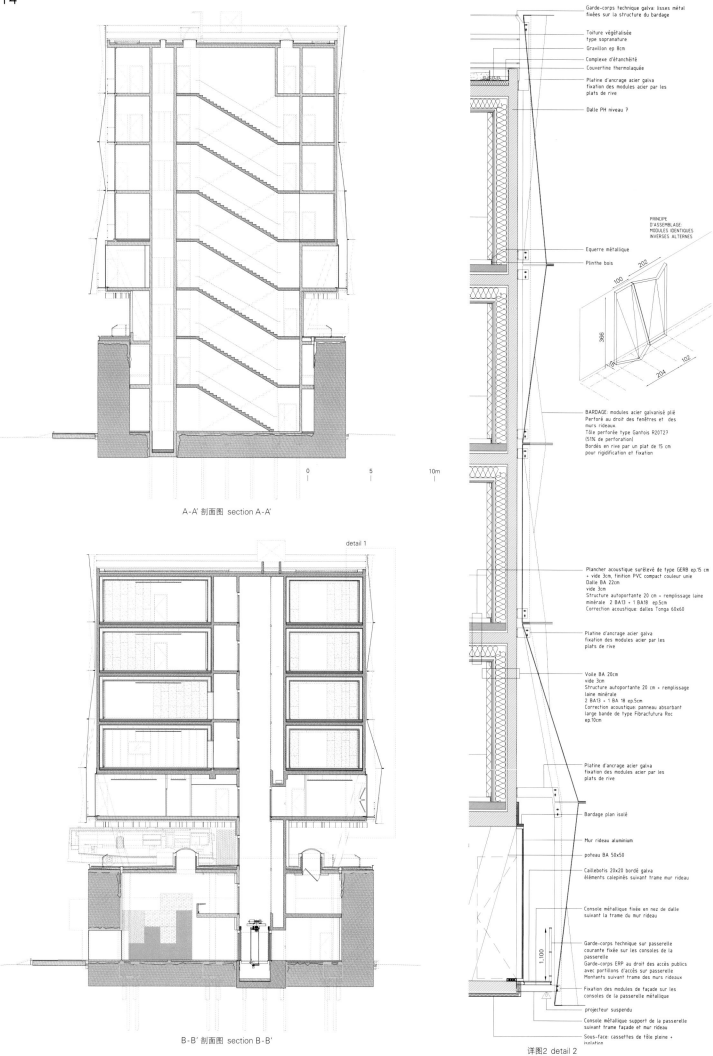

A-A' 剖面图 section A-A'

B-B' 剖面图 section B-B'

详图2 detail 2

格栅正面——烟控室
grid frontage _ smoke control room

详图1 detail 1

previous Dubigeon warehouse and near the Alstom warehouse (the upcoming center of the "Quartier de la Création") and the old shipyard turned into great gardens. This context provided a complex opportunity to gather the memories of a place abandoned by history to remind us that culture is a continuum made of sedimentation and fertile rediscovery. By gathering the public and artists around expressive or chamber concerts, significant projects and initiatives emerge: two buildings testify to the dual ambition of a unique project, with the architecture of TETRARC representing the identity of Nantes. The historical places are industrial warehouses, which include the techniques that built them to allow science and aesthetics to meet to create a new kind of beauty. The history of the ships is built on the strength of the skilled and talented workers. The history of the artists allowed their eyes to perceive this universe – from Monet, the painter of the brand new steel structures, to Becher, who used to forget metallic shapes. Finally, the history of the industrial production organization resulted in an experimental workshop, an assembly line, a rectification workshop, and an export site all being placed under one roof, creating a living picture of the concept of flexibility. The project comprises three distinct elements: the 400-people hall (with offices in it), the 1200-people hall – with a public hall connecting the two,

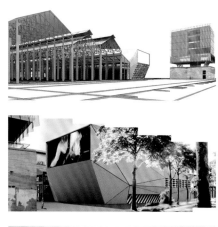

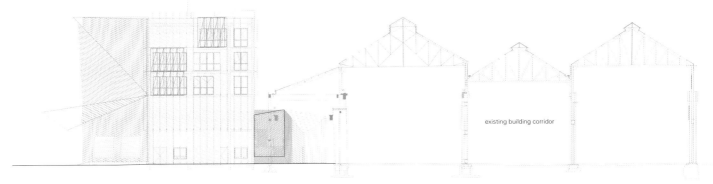
北立面 north elevation

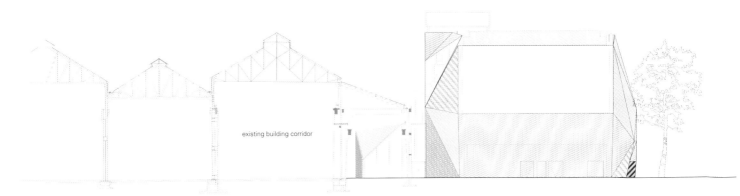
南立面 south elevation

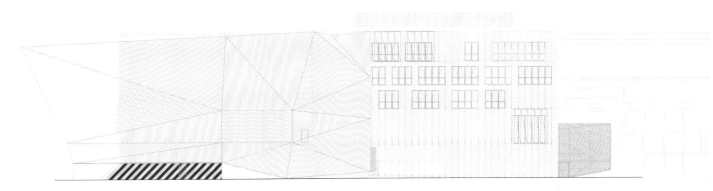
西立面 west elevation

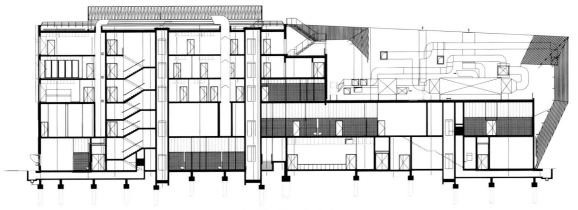
C-C' 剖面图 section C-C'

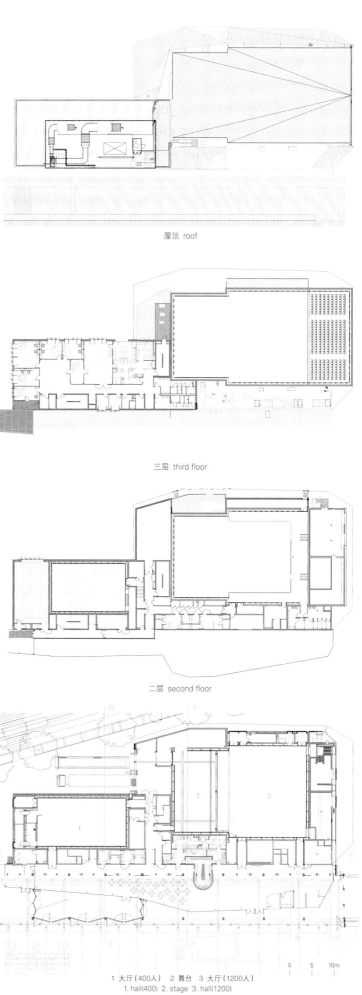

屋顶 roof

三层 third floor

二层 second floor

1 大厅(400人)　2 舞台　3 大厅(1200人)
1. hall(400)　2. stage　3. hall(1200)
一层 first floor

inserted in the concrete post frame of the Dubigeon warehouse – and finally the studios, located in a volume suspended above an old air-raid shelter.

Three moments in architecture

The noticeable space from Léon Bureau Boulevard and the signal at the intersection with Boulevard de la Prairie aux Ducs are dedicated to creating the immediately perceptible three major components of the program presented to the architects: a reception area, two showrooms, and a suite of offices and studios. Yet giving form to equipment allows the public to discover new worlds invented by the explorers of sounds and images that were meant to design an expressive and original work of architecture. The response is the metal plates that are folded, assembled, stacked, and combined freely to cover the functional interior space.

Warehouse connection

La Fabrique is adjacent to Dubigeon Warehouse, using the concrete columns of the latter's outgrowth to form a protected public space in which a help desk, AIM, product sales desk, and a bar/restaurant are located. This long, irregularly shaped hall is simply and brightly furnished, creating a smooth continuity with the halls that host the "Island of the Machine and Elephant". It also offers easy access to the largest mall outlet supplying the area. Soon, the construction of the National School of Architecture and Alstom warehouse will be transformed into the College of Fine Arts.

Maxi-Micro Hall

Large, metallic-covered panoramic windows open to the urban ocean, creating one part of the front view to Léon Bureau Boulevard, establishing a beacon calling the public together. The building is extended by beveled space, adjusted through long metal tapering plates, and punched by large doors with the size of semi-trailers. It is also dotted with perforations through which tubes and fittings can occasionally be seen, thereby revealing the complexity of the extensive equipment that enables 1,200 people to enjoy performances by famous artists in the Maxi Hall and for 200 inquisitive minds to discover emerging talents in the Micro Hall. If required, these performances are provided via the platform for sonic experiments to generate multimedia opportunities.

Engine of talents

From the concrete foundation of a hollow and leveled anti-aircraft shelter, a lifting device appears to hoist skyward, opening to a future of seven glittering floors of offices and recording/rehearsal studios presented with an urban function. Through this dynamic thrust, between the walls of history and the material of the bunker, La Place offers moments to be shared in small groups. Under colorful shaded lamps, the cultural cafe unites 100 people for mini-concerts and meetings. Above, La Terrace extends this reception with its MAO (Musique Assistée par Ordinateur) bus embedded into the building. TETRARC

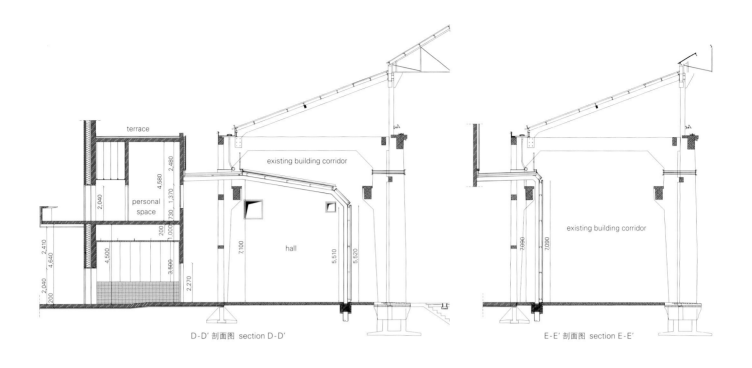

D-D' 剖面图 section D-D'　　　　　　E-E' 剖面图 section E-E'

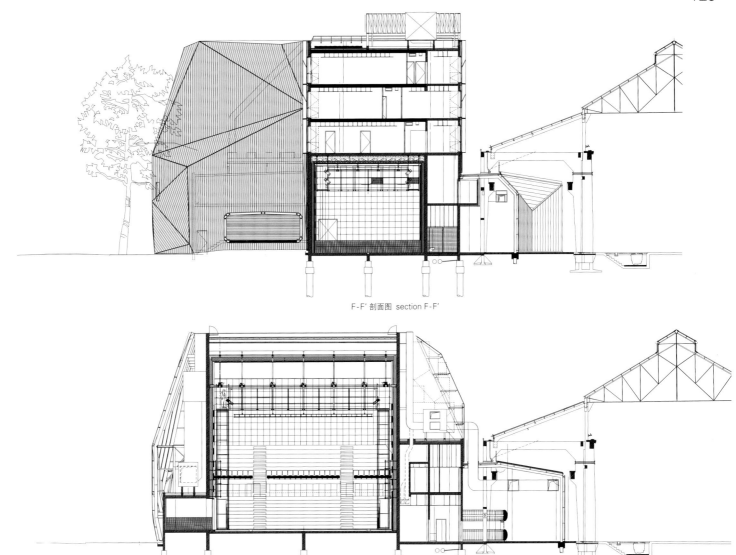

F-F' 剖面图 section F-F'

G-G' 剖面图 section G-G'

项目名称：La Fabrique
地点：Nantes, France
建筑师：TETRARC Architectes
项目经理：Michel Bertreux
项目主管：Rémy Tymen
项目研究：Guillaume Blanchard
经济学家：CMB
结构工程师：SERBA 流体工程师：AREA
HQE顾问工程师：Area canopée
音响师：Atelier rouch
甲方：Ville de Nantes,
SAMOA (Société d'Aménagement de la Métropole Ouest Atlantique)
建筑物表面：7,222m²
造价：16,2 M€ HT 设计时间：2005 竣工时间：2011
摄影师：©Stéphane Chalmeau

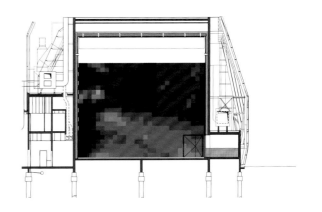

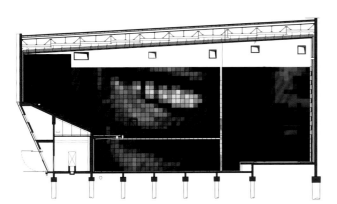

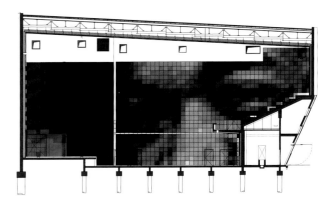

学校建筑演变为一种对基本的人类需求做出的形式反应。这种信息的传递范围从简单的几何学和交流方案延续到当代正规教育的复杂性——现代社会复杂性的一个基本构成要素。

在这一背景下，学校建筑因为本身是公共、政治和市民生活活动中的教育工具，包含了人类寻求文明中的一个关键元素。

对原有的教育基础设施进行增建或改建的特殊项目反映了这一活动本身的动态特征，也再现了我们作为社会性物种具有的某些基本特性的本质。对具有明显简易性特征的建筑（例如Alejandro de la Sota Martínez建造的Maravillas学校体育馆）进行增建，与之后论述的六个项目代表了缝合新旧结构之间的细微裂缝的一种富有启发性的方法，这种方法应用在以多样性与变化为自然特征的项目中。

教育基础设施的动态本质——一个真正的学校重建和不变的形式修建过程——最终将成为学校融入社会的一个永恒过程。

School architecture has evolved as a formal response to a basic human need. The transmission of information ranges from simple geometric and communicational schemes to the sophistication of formal contemporary education – a fundamental component of the complexity of modern societies.

Within this context, school architecture constitutes a key element in the quest for civility, being itself a pedagogic instrument for the exercise of public, political and civic life.

The particular project of adding and transforming pre-existent educational infrastructures reflects the dynamic character of the activity itself, and represents the nature of several basic traits of our social species. Adding on a monument that is paradoxically characterized by its apparent simplicity, a canonical example such as Alejandro de la Sota Martínez's gym for the Colegio Maravillas, together with the six projects that follow, represents an enlightening approach to the act of sewing delicate seams between the old and the new, within a program that is naturally signed by multiplicity and change.

The dynamic nature of educational infrastructure, a true palimpsest and constant formal revisions, is in the end a permanent process of integrating the school into the city.

D.Manuel I中学_ D.Manuel I Secondary School /BFJ Arquitectos
Vila Viçosa高级中学_Vila Viçosa High School/Cândido Chuva Gomes Arquitectos
Francisco de Arruda学校_Francisco de Arruda School / José Neves
拉滕贝格中学_Rattenberg Secondary School/Architekt Daniel Fügenschuh ZT GmbH
贝尔纳多特学校的扩建_Bernadotte School Extension/Tegnestuen Vandkunsten
裴斯塔洛齐学校_Pestalozzi School/SOMAA + Gabriele Dongus
对朴素的学校建筑进行增建_Adding on the Modest Monument / Jorge Alberto Mejía Hernández

对朴素的学校建筑进行增建
Adding on the Modest Monument

"没有建立实质结构的社会更倾向于在一些聚集点——池塘、林荫、火、伟大的教师——进行活动,其占据的空间的外部界限模糊,并可根据功能上的需求和罕见的常规性进行调节。"

——Reyner Banham[1]

人类聚集在一个可以进行知识传递的中心,并建立其文化地位。教育是最基本的社会活动,其功能是集中人们分散的思想,允许我们(至少是希望我们)理解自然存在的神秘。

从男人和女人最初围坐在老师身边,到古典文化的重要学术环境;从那时到现在,想要解决基本人类问题的建筑形状发生持续变化。这种变化在许多方面中最关键的两个方面得到明显体现。

首先是对活动本身的处理。不同的教学方法、阶段和技术可以对聚集地、相互作用或对话、集中或分散、隔离外部干扰或关注的世界扩展进行阐释。

其次,教育规划的进展根据需求及时调整,以及时适应种种变化,对学校进行真正的重建,即是决定建筑本质的复杂性表现。人们可以推断出,建筑不仅仅是一种特殊的主题,或者众多教学方法和逐渐引导学术活动的学校之一,它凭借自身的实力成为教学实体,成为坚固的信息储藏室,在复杂性方面具有可传递性和可学习性。

或许教育规划项目在建筑原则所要求的项目中是最为复杂的一种。它的形式深深地植根于政治和哲学环境中,并且在建筑类型学的严格限制和传统的城市限制条件内体现出来。对于一座学校来说,其最突出之处——然而更是其朴实之处——在于其所包含的深厚的社会内涵,而这是源于教育是一件宝贵的工具这一事实。

从这种意义上说,学校组成了最市民化的公共活动的一部分;体现了具有公民权的公共机构,代表了一个独有的社会所引以为傲的(所以也被认为是值得分享和保留的)和最重要的事情:它的目的所在。最虚幻的乌托邦建筑(远非近来常见的技术和形式方面的突发奇想)依靠真正的乐观主义发挥作用。学校建筑通常将其抱负延伸至未来;将其大胆

"…societies which do not build substantial structures tend to group their activities around some central focus – a water hole, a shade tree, a fire, a great teacher – and inhabit a space whose external boundaries are vague, adjustable, according to functional need and rarely regularity."– Reyner Banham[1]

Gathered around a central focus that allows for the transmission of knowledge, humanity exercises its cultural status. Education is the fundamental social activity, the function concentrates people around the abstraction of ideas and allows us to (at least hope to) understand the mysteries of natural existence.

From the initial circle of men and women surrounding their teacher, to the monumental academia of classic culture; and from then to the present, the architectural repertoire of shapes that intend to solve this basic human problem constantly varies. It conspicuously does so in two crucial aspects, among many.

The first deals with the activity itself. Different pedagogic approaches, stages and techniques account for the implementation of privilege focus, interaction or dialogue; concentration or dispersion; isolation from external disturbance or expansion into a world to be observed.

Secondly, the evolution of the educational program has by necessity adapted and adjusted itself to a myriad changes in time, creating a true palimpsest, an expression of complexity that necessarily determines the nature of the architecture. More than a particular subject, then, or one among the many teaching methods and schools that generally guide academic activity, it is possible to assume architecture as a pedagogic entity in its own right, being a solid repository of information, transmissible and therefore learnable, on the subject of complexity.

The educational program is, perhaps, one of the most sophisticated among the projects requested from the architectural discipline. Its form is deep rooted in political and philosophical considerations of condition, and embodied within the strict limits of architectural typology and conventional urban constraints. A monument par excellence – however modest, even – a school involves profound social connotations, given the fact that education is usually treasured as a precious tool.

In this sense, schools constitute what is perhaps the most civic of public activities; they embody the institution in which citizenship is acquired, represent what a particular society is most proud of (and therefore considered to worth sharing or preserving), and most important: what it aims for. The most visionary and utopian architectures (far from technological and formal boutades, so common these days) operate on genuine optimism. Schools usually extend their aspirations into the future; their boldness is often hidden within the apparently prosaic classrooms or simple halls.

由Alejandro de la Sota设计的位于马德里的Maravillas学校的体育馆扩建项目，1962年
Gymnasium extension of Maravillas College in Madrid by Alejandro de la Sota, 1962

D.Manuel I中学的现代化
D. Manuel I Secondary School's modernization

设计隐藏于表面平凡的教室和简易的大厅中。

由Alejandro de la Sota设计，位于马德里的Maravillas学校扩建项目明确地体现了简易性中所蕴涵的复杂性，是真正名副其实的奇观！

最初的设计概要于1959年由Alfredo Ramón-Laca[2]提出，主要是为了弥合瓜达基维尔河和Juan Costa大街之间12m的高度差。后来，由于该项目需要更大的空间建立体育馆，建筑师决定在斜坡上嵌入一个体量，使之抬高，位于先前存在的设施（带有一个耐用的操场）的水平高度。

De la Sota在1962年设计的插入结构正是完美地利用了之前的建筑，对最初的整体设计方案进行了重要修整。他的提议仍被看作是复杂的分层式策略，通过大胆地诠释技术方案，使其作为空间和功能方面的机遇。由于体育馆使自由的设计变得具有强制性，从建造初期开始，单一立面上的木柱和嵌入区域的后墙都应支撑比正常标准更大的桁架。因此整个屋顶表面都被看作有用的内部空间，并在体育馆和操场之间引入了起媒介作用的楼层，用于建造一种特殊的教室（倾斜在倒置的桁架前部）。一个带盖子的巨大盒状建筑（1959年）成为一座四层建筑（1962年），它包括一间半地下室（容纳一个洞穴般的池塘空间）、一个建有体育馆的基础地面、建有教学设施并发挥媒介作用的楼层以及另一个基础地面，或者一个屋顶，还包含原有建筑操场。

项目的水平分层在涉及到位置的具体问题的纵向解决方案中得以体现。由于扩建的结构带有一个单一的立面，并且这一单一立面面向最繁华的街道，楔形的厚平板教室易受到外部噪音的影响。De la Sota在垂直方向引入一层额外的空置空间，通过双层玻璃来照亮房间，玻璃使得外部的光线透过体育馆和小礼堂唯一的窗户进入室内，为要求专注力的区域创建屏障，同时将较低处的喧闹区域隔离在室外。

之前的扩建工程（原始的建筑+最初的项目+最终的扩建）的扩建部分在具体建筑的第三级发展中或许能够解释整个项目的奇迹。当建筑师在简单的甚至是平常的问题中坚持某些决定时（很明显，即使在最繁荣的社会，教学建筑也几乎不允许建设得过于奢华，或者大胆、无节制地使用材料），它最终就成为建筑力量的一次真正教育。这种力量体现了20世纪下半叶建筑文化所提出的复杂条件。

A conspicuous show of such complexity within simplicity is the school extension designed by Alejandro de la Sota for the Colegio Maravillas, in Madrid, a true wonder – justice to the name!

The initial brief developed in 1959 by Alfredo Ramón-Laca[2] fundamentally dealt with a 12-meter difference between Guadalquivir and Juan Costa streets. Since the program asked for a big space to be assigned gym functions, the architect decided to insert a volume within the slope, topping it at the level of the pre-existent facilities with a hard playground.

De la Sota's 1962 intervention took on from this *parti*, elaborating substantially on the initial scheme. His proposal is still regarded as a sophisticated layering strategy, by means of a bold reading of technical solutions as both spatial and functional opportunities. Since the gym makes a free plan mandatory, it was clear from the start that posts located on the single facade and the back wall of the excavation should support larger-than-normal trusses. Thus, the whole section of the roof surface was considered as useful interior space in itself, introducing an intermediate floor between the playground and the gym, used for a very particular type of classroom (sloped within the bellies of these inverted trusses). A huge box with a lid (1959) became a four-story building (1962) comprising a semi-sunken basement containing a cave-like pool space, a base ground with a gym, an intermediate floor with teaching facilities, and another base ground, or a roof, with playgrounds for the existing buildings.

The horizontal layering of this distribution is also presented in the vertical solution of specific problems concerning location. Since the extension has a single elevation, and this sole facade is open to the busiest street, the wedge-like, thick slab classrooms appeared vulnerable to the outside noise. De la Sota then introduced an additional layer of empty space in the vertical, illuminating the rooms through a double-skin glass that allows intake of external light through a single window for both the gym and the small auditorium, creating insulation for areas that require concentration while keeping the lower, noisier area conveniently closed to the outside.

The addition to a previous extension project (original building + initial project + final intervention), operating at the third degree of evolution of a particular architecture, might explain the miracle of the whole project. While decisions are kept within the realm of the simple, even prosaic problem solving concerns of the architect (and it must be clear that even in the most prosperous soci-

照片提供: José Neves
(©Laura Castro Caldas & Paulo Cintra)

Francisco de Arruda学校, 首建于1956年, 原始建筑进行了修复, 并且还增建了三种新元素
Francisco de Arruda School, first built in 1956. The original buildings were restored and three new elements were added.

这一允许教学建筑超越传统学术项目并突出其复杂性现象的代表性特征, 促使了社区市民对学校给予资金支出。与其他的公共建筑相反, 无论是宗教建筑还是政府建筑, 普通大众都似乎与组织教育的机构和实践教育的地方有着极为密切的关系。不管学校具有怎样的传统性或试验性, 是如何令人惊讶或朴素, 一方面, 它通常都被看作是敬意和赞美的混合体, 另一方面则是希望与感激相交织。

当对教学建筑替代物的试验源自想要对传统学校中传统元素的形状和分布做出修改时 (从六边形蜂窝状到多焦点的环境), 大多数建筑师很少能够突破邻近的教室常用的增建策略的限制。

在传统和改建学校里, 经常能见证超凡的功能灵活性。在常规的工作日, 学生可以使用部分建筑, 而在周末、工作日的晚上以及像选举日和节日这样的特殊日子时, 整个社区都能受益于这种基础设施的集中性特征。从体育馆、大厅、礼堂和开放空间举行的社区会议, 到教室里举办的成人教育讲习班, 当代的许多学校除了基础功能, 还具有活跃和有用的功能, 几乎可以随时使用。

在下面介绍的原有学校的扩建案例中, 不同程度的整修项目都试图将七个案例带入更高的情景化水平, 或者通过某段要求进行新活动的时期, 或者在城市环境 (由学校自身的再定位形成) 的显著变化中能表现自身张力的场所来实现。

以下项目中有三个是作为葡萄牙学校现代化项目的一部分, 由Parque Escolar EPE筹划, 多方同时进行, 而不只限于普通客户。

由贝贾的BFJ建筑师事务所设计的D.Manuel I中学也是上文提到的复杂性处理方面的范例。BFJ事务所设计的增建结构以严格的功能规划为基础, 这项规划将已建区和未建区之间的场地分割成具有显著使用特色的相关区域; 把先前存在的各片区域连成一处无缝的薄纱状区域; 这个增建结构引入新的设施, 既能提供永久性服务, 又能为将来的转变积聚相关功能。

Vila Viçosa高级中学由Cândido Chuva Gomes建筑师事务所设计, 是学校的再建工程。建筑师将先前的增建结构中破落的薄纱状结构拆除, 然后将其余松散的部分与全新的建筑重新连接在一起, 新建筑既具

eties the educational program seldom allows for luxury, material boldness or excess), the overall result is a true lesson of architectural power which reveals the complex conditions proposed by architectural culture in the second part of the 20th century.

This representational character, that allows educational architecture to highlight the phenomenon of complexity beyond conventional academic subjects, leads to the appropriation of schools by the citizens of the community they belong to. Contrary to other public buildings, be they religious or governmental, the common man and woman appear to have an extremely close relation with the institution of organized education and the place where it happens. No matter how traditional or experimental, how astonishing or modest, the school is usually seen with a mix or respect and admiration on the one hand, hope and gratitude on the other.

While experiments with pedagogic alternatives have derived in original proposals that intend to modify the shape and distribution of the traditional elements of the conventional school (from hexagonal beehives to multi-focal environments), most architects seldom transcend the limits of the habitual additive strategy of contiguous classrooms.

In both traditional and alternative schools though, it is not unusual to witness extraordinary functional flexibility, allowing for the use of part of the building by students on regular weekdays, while the community as a whole benefits from the more collective features of the infrastructure on weekends, weeknights and under exceptional circumstances such as election days or festive events. From community meetings held in gyms, halls, auditorium and open spaces, to adult education workshops held in classrooms, many contemporary schools remain active and useful beyond their basic functions, on a close to 24/7 basis.

In the case of the following additions to pre-existing schools, it is clear that the project of refurbishment, at different levels, tries to bring these seven examples to a higher level of contextualization, either with a time that asks for new activities, or with a place that expresses its own tension in the dramatic transformations of the urban context, proposed through the re-orientation of the school itself.

Three of the following projects have been promoted by the Parque Escolar-EPE, as part of the Portugese School Modernization Program, and coincide in several aspects, beyond the common client.

The D. Manuel I Secondary School by BFJ Arquitectos in Beja con-

贝尔纳多特学校的扩建项目，增建了一座图书馆和一座带有舞台设施和室外空间的体育馆
Bernadotte School extension, adding a library and a gym with stage facilities, outdoor space

照片提供：Tegnestuen Vandkunsten (©Adam Mørk)

位于前奥古斯丁修道院和一座体育馆之间的拉滕贝格中学的扩建项目
Extension to Rattenberg Secondary School, between a former Augustine monastery and a gym

照片提供：Daniel Fugenschuh (©Christian Flatscher)

有所需的功能，又重塑了学校本身的威严形象。

由José Neves设计的里斯本Francisco de Arruda学校本身谨慎地（建筑师用"悄悄地"这一词来形容）定位为先前存在的建筑所组成的"小城市"中的必要的增建结构。仅仅完成天井的设计之后，建筑师在这种环境中所融入的质朴感与敬意就产生了极为优雅的效果。

考虑到建筑体量的大小变化范围，下文有三个项目建在了纬度更高的地方。

其中的一个项目是由Tegnestuen Vandkunsten设计的贝尔纳多特学校的扩建项目。项目位于哥本哈根，这所小型增建学校包括几栋陈旧的别墅，看起来如同迷宫一样，具有鲜明的特色，但缺乏实用的空间。这一小型扩建项目包括一座图书馆和一个小礼堂（体育馆和舞台），它保留了学校的迷宫般的复杂性，同时用更统一的手法彻底简化整体形象，使扩建的项目处于平衡的状态。在平面和剖面中，建筑师都采用L形体块，通过这一精致的嵌入结构使新旧建筑统一起来，并通过黑色金属的中性物质性以及藤蔓植物（爬满扩建物和别墅的立面）覆盖的钢框架来完善建筑。

总部位于斯图加特的SOMAA事务所也规划了一个精致的嵌入结构，对莱昂贝格小镇（德国）的裴斯塔洛齐学校进行了扩建。新的建筑将为采用非传统学习方式的孩子们所用，它在旧街区中确定出一个庭院，并且在封闭空间（外部）和开放空间（针对旧的设施）之间，粗糙的混凝土外部和平滑的紫罗兰色室内之间创造强烈的对比，这种平衡具有张力，在剖面的多标量本质（半地下室，在城市范围内将其调整为儿童体格大小）中体现出来。他们仔细地设计了呈对角的几何形框架，用来浇筑混凝土，以完善建筑效果，创建可以摇摆的体量印象，而这个体量几乎倚靠在一个小斜坡上。

由Daniel Fügenschuh设计的奥地利拉滕贝格中学，在薄纱状结构中嵌入了其设计师称之为"国家中最小的小镇"的结构。这一具有意义的嵌入结构位于古老的奥古斯丁修道院旁（其本身也是一种重建结构，建筑表皮为哥特式和巴洛克风格），为整个小镇重建了公共空间，它也证明了严肃和真诚使得新旧建筑无需在风格上做出妥协也能联系起来。一种

stitutes one of the aforementioned operations within complexity. Based on a rigorous functional scheme that divides the site between built and free areas, into clearly characterized use-related zones, and while sewing the pre-existent pieces into a tighter tissue, BFJ's intervention introduces new equipments that serve both permanent and articulating functions for even further transformations.

Vila Viçosa High School by Cândido Chuva Gomes Arquitectos is a reconstitution of the palimpsest. Cutting off decaying tissue from previous interventions, the remaining loose parts are tied together again with brand new buildings that both fulfill necessary functions but also re-dignify the image of the school itself.

The project for the Francisco de Arruda School, in Lisbon, by José Neves, assumes its role discretely ("quietly", say the architects) as a necessary intervention within a "small city", constituted by the pre-existing buildings. By simply completing the figure of the patio, the modesty and respect with which the architects intervene in such a context produce extremely elegant results.

On the side of the spectrum, given their size, there comes three projects from higher latitudes.

One of them is the Bernadotte School Extension by Tegnestuen Vandkunsten. Located in Copenhagen, the small alternative school housed in a couple of old villas appeared labyrinthine, impregnated with a strong character but lacking in actual useful space. The extension to the minute program, which includes a library and a small aula (gym + stage) preserves the complexity of the labyrinthine quality of the school, while balancing it by means of a radical simplification of the overall image in more unitary terms. Using a L shaped block, both in plan and section, the architects intend to unify the old and new through a subtle intervention, complete with the neutral materiality of black metal and steel tensors that are expected to be covered by vines, creeping over the facades of extension and villas alike.

Stuttgart based SOMAA propose a subtle intervention too, with their addition to the Pestalozzi School in the small town of Leonberg (Germany). The new block, to be used by children confronted with alternative learning methods, defines a courtyard within the old neighborhood. Strong contrasts are created between closed (outside) and open spaces (to the older facilities), as well as between a rough concrete exterior and a smooth violet interior, the balance is tense, and is also presented in the multi-scalar nature of the section (semi-sunken, scaled to the size of children at an urban scale). The carefully designed diagonal geometry of the formwork used to pour the concrete completes the effect, creating the im-

《雅典学校》，Sanzio Raffaello绘制，1509—1510年
School of Athens, Sanzio Raffaello, 1509~1510

建筑师赫尔曼·赫兹伯格建造的代尔夫特蒙特梭利学校，1960—1966年
Delft Montessori School, Architect Herman Hertzberger, 1960~1966

显然全新的、不加修饰的现代建筑形式融入了城市古老的华丽装饰中。

这就是学校改建和扩建项目：精致、紧张、平衡而复杂；同时又朴素而具有纪念性。我必须强调，它全部始于似乎有话要说的某个人，或者是洞察力和经验。它也始于其他围绕着最简单的几何理念的事物。从那时起，学校成为一座小屋，本质独立；或者是一个球场，融入到高密度的城市环境中。它可以是Raphael所描绘的《雅典学校》，是赫兹伯格建立的蒙特梭利学校的具体化，或者理查德·诺伊特拉设计的波多黎各帝国委员会，亦或是阿纳·雅各布森熠熠生辉的Munkegard建筑，甚至是贫穷社区外围的一个增建结构，如Obranegra设计的麦德林圣多明哥学校——所有这些建筑都能够成为学校。但无论它最初是什么，有一点值得肯定的是：它最有可能不断地改变。

学校建筑的内部具有柔和性，满是书和书包、笔记本和纸；与一个短暂的使用者的经历——其由多个个体构成的结构在永久的迁移过程中被一代代人们的分散而分解——作为社会中的学校角色一直在改变。

学校扩建正代表着这种运动，即这种个体能够顺利通过但是机构仍在坚持的知识积累。加、减、乘、除全都运用在算术这种有着最独特特性的人类知识上。建筑通过在持续的变化中满足人类的需求来进行操作，并随着每一次活动，在具有动态特征的生活中教育我们所有人。

当这些建筑获得成功的时候——许多学校的确应被赋予这样的资格，远比其他相关建筑更加光彩熠熠、闻名遐迩——这些建筑在单个项目和老项目后来的改造中具体化；这些增建结构使得如此多的生活在其中得以展现；最后构成终极设计的核心：城市——最大的学校。

1. *The Architecture of the Well Tempered Environment*, Banham Reyner, London: the Architectural Press, 1969, p.20
2. http://urbancidades.wordpress.com/2010/01/15/gimnasio-del-colegio-maravillas-de-madrid-1962/, accessed 03/11/2012

pression of a vibrating volume, almost leaning over a slight slope. Rattenberg Secondary School, Austria, by Daniel Fügenschuch, intervenes within the delicate tissue of what is described by its author as "the smallest town in the country". Beside an old Augustine monastery (a palimpsest itself, bearing Gothic and Baroque scars on its skin), and recreating public space for the whole township, this no-nonsense intervention proves that sobriety and sincerity allow the new and the old to relate without the need of appealing to stylistic concessions. An evidently new, unashamedly contemporary form blends comfortably in the old urban filigree. Such is the project of school extensions and additions: subtle, tense, balanced and complex; modest and monumental at the same time. It all starts – I must insist – with someone who appears to have something to say; some insight, some experience. It starts with others too, surrounding the idea in the simplest of geometries. From then on, the school becomes a hut, freestanding in nature; or a court, integrated within the city's density. It can be the *School of Athens*, as depicted by Raphael, the materialization of Montessori School by Hertzberger, Richard Neutra's Imperial Commissions in Puerto Rico; or Arne Jacobsen's brilliant Munkegard; even a periphery intervention in a poor neighborhood, such as Obranegra's Santo Domingo school in Medellín – all of this, it can be. But whatever its initial reality is, one thing is certain: it will most probably be in constant change.

The softness of an interior full of books and bags, notebooks and paper, together with the testimony of an ephemeral user – the body made of many individuals diluted within the abstraction of generations in perpetual flux – is as ever-changing as the school's role in society.

School extensions represent this movement, this accumulation of knowledge that allows individuals to pass while institutions persist. Additions, subtractions, divisions and multiplications, all are operated on the arithmetic of human knowledge, our most distinct and particular trait. With every movement, architecture educates all of us in the dynamic life, by framing the need to operate in the persistence of change.

When successful – and many schools really deserve this qualification, beyond their more photogenic and famous relatives – these architectures are crystallized in individual projects and posterior transformations of older buildings; these interventions allow so much life to happen within their entrails, and end up constituting the kernel of the ultimate project: the city – the biggest school of all. Jorge A. Mejía Hernández

D.Manuel I中学

BFJ Arquitectos

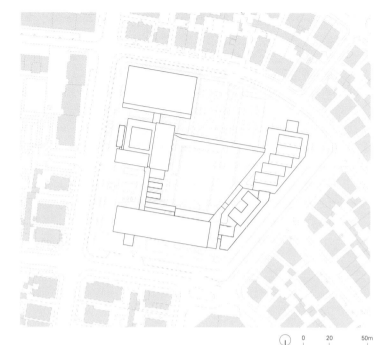

照片提供: the architect

贝贾的D.Manuel I中学的更新改造是Parque Escolar-EPE推动的中学现代化规划一期的一部分，可以总结为以下措施：

——修复原有建筑是为了增强其空间和构造特色，并使其符合时下对舒适性、安全性和便利度的要求。

——新建的一座实验大楼嵌入地形中，将原有的主体建筑和旧车间联系起来。

——原有建筑的修复还包括学生的休息区。休息区因其所处的位置而被构想为多个原有建筑之间的结合点，创建出一个多功能的带顶区域，并促进整个学校社团之间的交流。

项目包含一个北向的不规则锯齿形屋顶，为室内空间提供了从各模块的不同坡面产生的发散的、变化的自然光，同时保护室内空间免受南向阳光的过度辐射。

空间完全采用自然通风，它是通过巧妙地安装在踢脚板上的进气口和拱顶高度上的可调节洞口来实现的，并且可以产生"烟囱效应"，来确保空间有正确的通风方式。

新图书馆和资源中心进行了整合，毗邻带顶的休息区，由于这片区域交通量大，并且与中央空间相连，因此被看做是学校的活力区。拥有了这个非常显著的位置，图书馆成为学校的互动交流区。

建筑的布局意在强调带顶休息区作为入口的重要性，为非正式的阅读区和多媒体教室提供便捷的通道。文献资料参阅区位于对面，是一处更封闭且更安静的区域。

——新的带顶体育场的建造包括运动及附属设施。它的选址和设计的目标是使建筑不受侧风和直射阳光的损害。淋浴间和更衣室有效利用了该地的自然坡度，两者很轻松地整合在一起，并获得了热舒适度。

——对建筑外部带顶的走廊进行扩建，使得原有建筑和新建筑能够密切联系和衔接。这种介入结构使人们可以在所有学校建筑里穿行而免受大雨和过度的太阳光带来的影响。

——通过规划多元化的核心和其各自的入口，学校向社区开放，以使各区域可以进行独立管理。这样，在多功能教室、图书馆、体育活动场等区域为学校外的社区提供独立的使用区将成为可能。

——将学校理解为一个全天候的学习空间，突破了正规教室的限制。学校空间被认为是用来在整个在校群体（例如学生、教师、其他人员、家长）内创造可以激发互动的环境条件，同时可以为个人和工作小组的发展创造"专门空间"，以便充实并拓展这个所谓的非正式的学习环境。

D.Manuel I Secondary School

The modernization of the D. Manuel I Secondary School in Beja is part of the first phase of the secondary school modernization programme promoted by Parque Escolar - EPE, and can be summarized in the following measures:

- Renovation of the existing buildings in order to enhance their spatial and constructive features in accordance with the most recent comfort, safety and accessibility requirements.
- Construction of a new laboratory building, inserted in the terrain to connect the main existing building with the old workshops.
- Renovation of the former covered student's recess area. Due to its location, it is conceived as a junction between the various existing buildings, creating a multifunctional covered area, encouraging social interaction amongst the entire school community. The project incorporates a north-facing irregular saw-tooth roof, providing this space with diffuse and variable natural light originating from the modules' different slopes, and simultaneously protecting the space from the excessive southern solar radiation. This space's ventilation is exclusively natural, attained by means of air intakes, strategically placed at the baseboard level and with adjustable openings at soffit height, allowing for the "chimney ef-

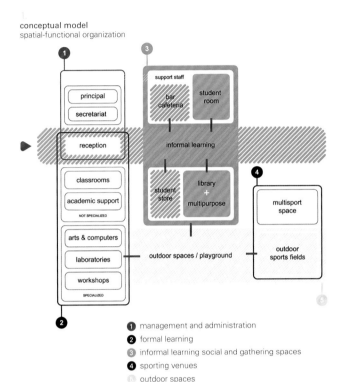

1.
conceptual model
spatial-functional organization

① management and administration
② formal learning
③ informal learning social and gathering spaces
④ sporting venues
⑤ outdoor spaces

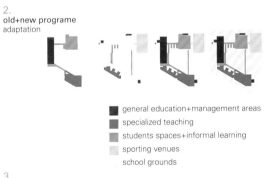

2.
old+new programe
adaptation

general education+management areas
specialized teaching
students spaces+informal learning
sporting venues
school grounds

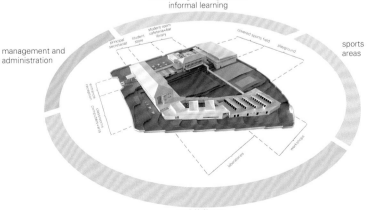

3.
proposal
division of the functional areas

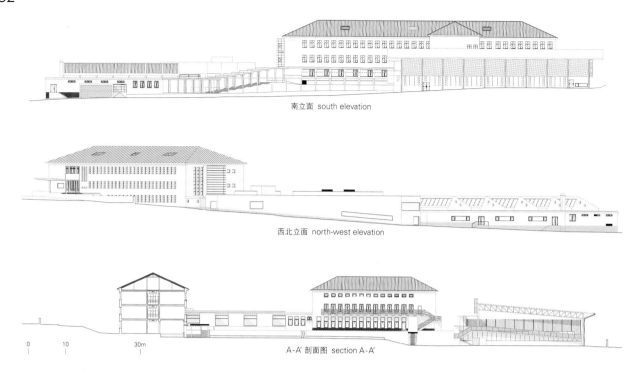

南立面 south elevation

西北立面 north-west elevation

A-A' 剖面图 section A-A'

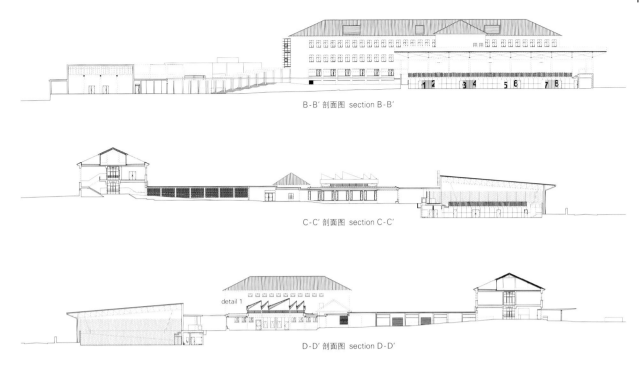

B-B' 剖面图 section B-B'

C-C' 剖面图 section C-C'

D-D' 剖面图 section D-D'

地下一层 first floor below ground

1 储藏室 2 工作间 3 车间 4 教室 5 实验室
6 电气设备室 7 制备室 8 技术区
1. storage 2. workroom 3. workshop 4. classroom 5. laboratory
6. electrical installations 7. preparation room 8. technical area

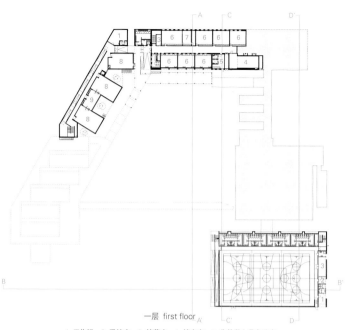

一层 first floor

1 工作间 2 看护室 3 储藏室 4 档案室 5 非教学人员办公室
6 教室 7 计算机车间 8 实验室 9 制备室
1. workroom 2. nurse 3. storage 4. archives 5. non-teaching staff room
6. classroom 7. computer workshop 8. laboratory 9. preparation room

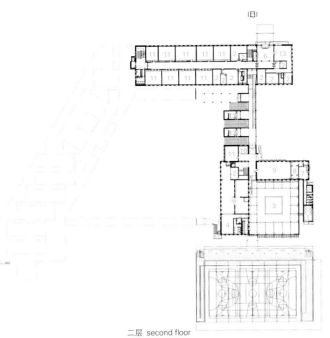
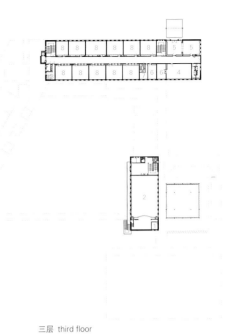

二层 second floor
1 技术区 2 工作间 3 带顶休息区 4 非正式学习区 5 图书馆 6 中庭 7 烹调区
8 厨房 9 自助餐厅/酒吧 10 学生会 11 教室 12 学生商店 13 秘书处
1. technical area 2. workroom 3. covered recess area 4. informal studying
5. library 6. atrium 7. cooking area 8. kitchen 9. cafeteria/bar
10. student association 11. classroom 12. student store 13. secretariat

三层 third floor
1 储藏室 2 多功能教室 3 技术区 4 教师工作间
5 教员室 6 会议室 7 工作间 8 教室
1. storage 2. multipurpose room 3. technical area 4. teacher workroom
5. teacher room 6. meeting room 7. workroom 8. classroom

项目名称：D.Manuel I Secondary School in Beja
地点：RuaSão João de Deus, Beja
建筑师：Francisco Amaral Pólvora, Bernardo Campos Pereira, José Amaral Pólvora
项目团队：Pedro Viana, Pedro Dorges, Diogo Andrade e Sousa, Nuno Lucas, Mariana Baptista, Júlio Senra, Ana Rita Oliveira, Pedro Prata
结构工程师：A2P Consult Estudos e Projectos
电气工程师：Energia Técnica 水力和天然气：Termifrio
集成安全：Espaço, Tempo e Utopia
空气调节：José Galvão Teles Engenheiros
能源认证：OPS Engenharia
声音处理：OPS Engenharia 固体废物处理：OPS Engenharia
景观建筑师：Nélia Martins e João Junqueira
总承包商：Mota-Engil
甲方：Parque Escolar-EPE
建筑面积：11,000m² 造价：EUR 7,850,000
设计时间：2007—2008 施工时间：2008—2009
摄影师：©FG+SG – Fotografia de Arquitectura (except as noted)

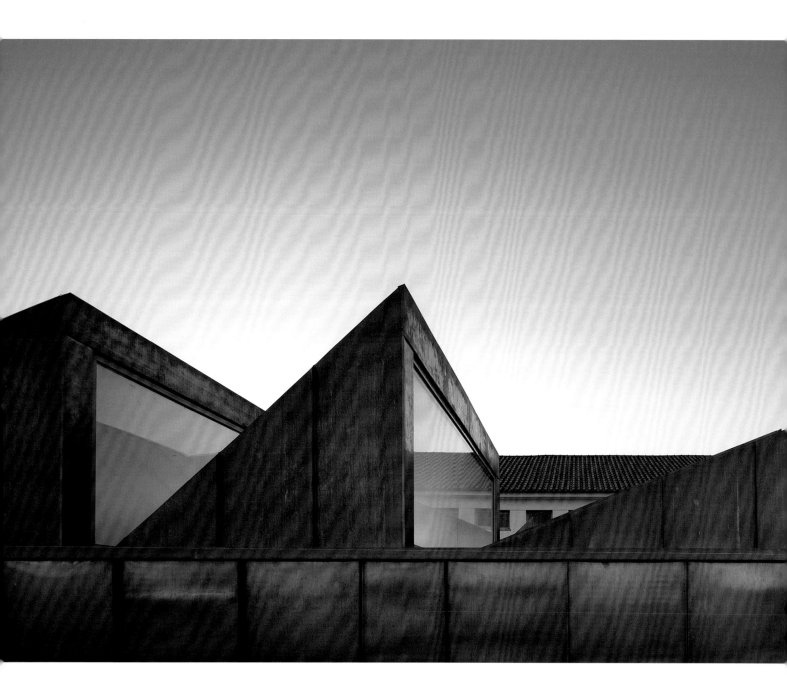

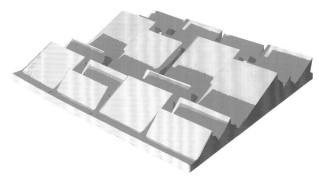

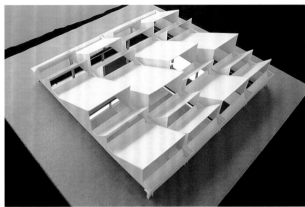

1. "vmzinc" cladding in titanium zinc n.º12
2. separation screen
3. extruded poliestyrene (e=50mm)
4. vapour barrier
5. maritime plywood support (e=22mm)
6. secondary structure in rhs50 sections
7. plasterboard (e=12.5mm)
8. zinc duct
9. window frame with glass slides adjustable opening "beta"
10. proofing system with pvc tiles type "sikaplan"
11. thermal slabs type "grisol"
12. groundsill in granite stone
13. double glass "saint-gobain": ext. "sgg securit planitherm" e=8mm; air box e=12mm; int. "sgg stadip 55.1"=10.4mm
14. fixed window in aluminum

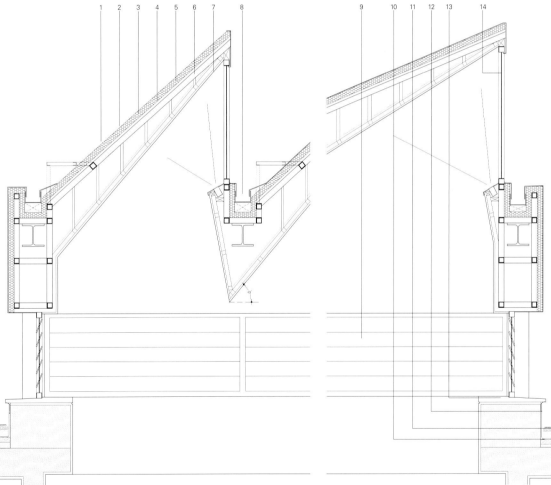

详图1 detail 1

fect" to assure correct space ventilation.

Integration of a new Library and Resource Centre: situated next to the covered recess area, this zone is conceived as the dynamic element in the school thanks to the high traffic it generates and its connection to the central space. Through its highly visible location the library becomes the interactive element of the school. The layout reveals the intent of emphasizing the importance of its entrance from the covered recess area, giving immediate access to the informal reading area and the multimedia room. The documentation reference area is situated on the opposite side, where a more protected and quiet space is ensured.

- Construction of a new covered sports field, including athletic and ancillary facilities. It is positioned and designed to offer protection from crosswinds and direct sunlight. The shower and change rooms take advantage of the terrain's natural slope, easing their integration and thermal comfort.

- Extension of the exterior covered galleries in order to intertwine and connect all of the existing and new buildings. With this intervention, it's possible to pass through all of the school buildings sheltered from the rain and the effects of excessive sunlight.

- Opening up the school to the community by means of organizing various nuclei and respective accesses, in order to allow for independent management of each zone. In this way, it is possible to provide independent zones for use to the community beyond the school in areas such as the multi-purpose room, the library, the sports areas, etc.

- Understanding the school as a full-time learning space, beyond the limits of the formal learning classroom. The school space is thought to create the conditions which activate interaction within the entire school community (i.e. students, teachers, other staff, parents), simultaneously allowing for the creation of "niches" for the development of individual and team work, so as to enrich and promote a so-called informal learning environment. BFJ Arquitectos

Vila Viçosa高级中学
Cândido Chuva Gomes Arquitectos

这个嵌入性项目的场地包括一个宽阔的、荒芜的学校操场；原有建筑之间生长着三棵高大的松树。位于中央的建筑具有社交和管理功能，其他两座建筑容纳了教室和实验室，另一座建筑仍作为工作室，最后，这里还建有一座体育馆。过时的功能、毫无吸引力的外形、破碎的结构导致了该设施的提早老化，同时其温暖度与亲和力也显然达不到学校社区的标准。除了上述几点，该建筑的保温情况、音质效果、空气处理方面的问题以及建筑障碍物和室外空间的杂乱无序等问题都亟需解决。

这些问题促使了一个基于建筑提案的新学校设计理念的诞生。首先，建筑师选定了需要保留下来的建筑，并绘制出一张与众不同的最终建筑图纸。建筑师将教学建筑与体育馆都保留了下来，并通过建造新的体量，来将这二者之间的空间填满，使之成为一座具有连续性的建筑。

建筑师从建造入口和庄严的大厅开始，来对这座"新"学校展开重新布局。入口和大厅空间延伸至室外，与户外广场融为一体。室内要求呈现给人一种简洁明亮的景象，清晰的路径使得不同的操作实体之间的连接变得十分便捷。

在室外，坚固的砌砖覆层不仅对建筑起到了保护作用，而且减少了建筑的维护费用，同时还使整座建筑综合体具有了统一性。玻璃纤维混凝土 (GRC) 网格有助于控制光线强度，且使建筑的外表具有一致性，同时增强了原有体量的整体性。项目的整个周围区域也经过了重新设计，以加强户外活动，增加与室内空间的联系，为该建筑呈现出一种优化了的结构框架。

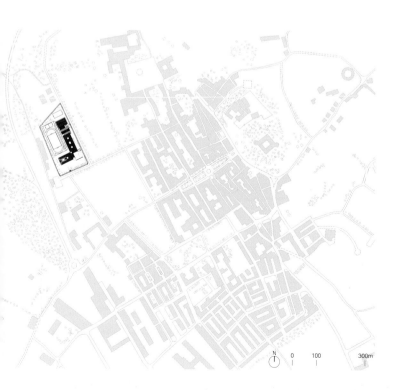

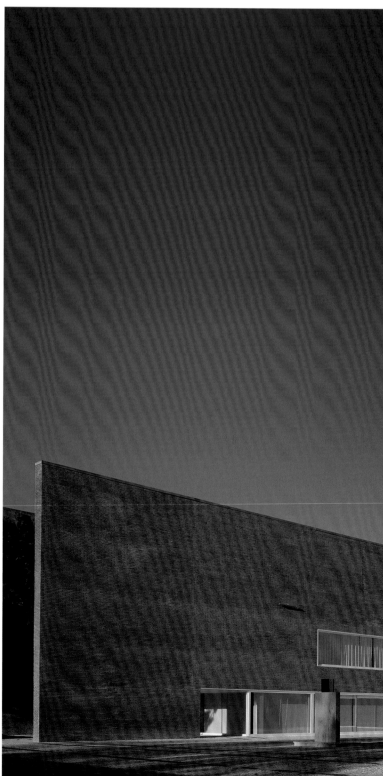

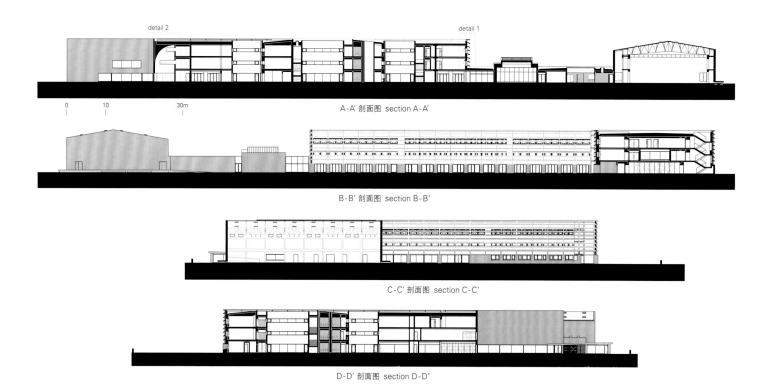

A-A' 剖面图 section A-A'

B-B' 剖面图 section B-B'

C-C' 剖面图 section C-C'

D-D' 剖面图 section D-D'

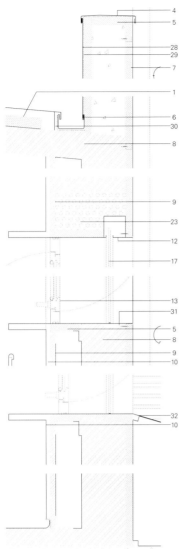
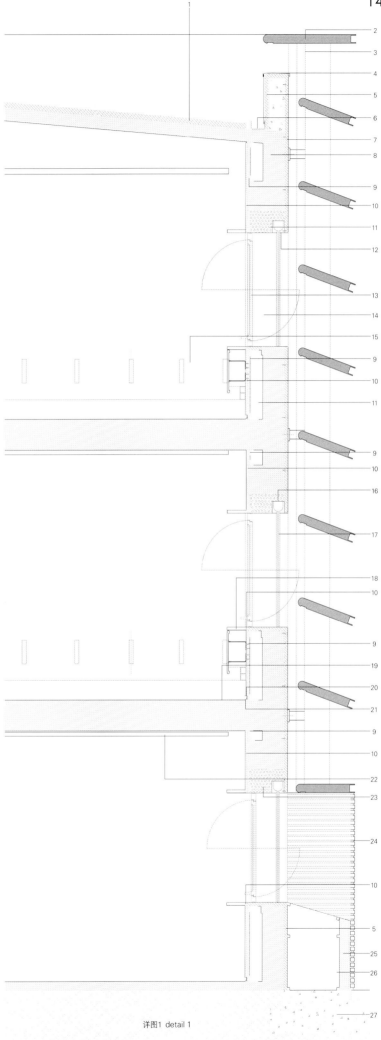

1. panels MT2 of "hiansa" or equivalent, fixed to the structure profile galvanized-filled spaces with insulation plates rigid extruded polystyrene foam with e=6cm, type "roofmate sl" or equivalent
2. gfrc panels "pavicentro" or equivalent
3. self-supporting structure for fixing the panels
4. rufus and clamps in zinc
5. thermal insulation plates in rigid extruded polystyrene foam with e=4cm, "wallmate" or equivalent
6. zinc gutter
7. thermal insulation system for outdoor type "weber.therm" or equivalent: thermal insulation boards 40mm extruded polystyrene
8. existing construction
9. panel composed of mineralized wood fibres, portland cement-bonded type "celenit n" 3cm or equivalent, fixed to the structure of I profiles in galvanized steel
10. plaster "seral" or equivalent
11. I-structure profiles in galvanized steel
12. box blinds bent sheet metal 2mm, metallized and painted white
13. double glazed sash of "maciça" or equivalent for enamelling
14. ring in sheet metal 2mm, metallized and painted
15. continuous element, for protecting walls and incorporating electrical trunking
16. profile I-galvanized steel for fixing the shutter box
17. roller blinds with manual crank "sombrol gc cabrio" or equivalent with supports in thermoplastic and side guides with 4mm cables in stainless steel, with tissue "t screen 5205" or equivalent of represtor
18. structure slatted iron tubes R4S 30x30 metallic lined with sheet iron bent, painted and metallic
19. floor "sikafloor 261 - autoalisante" or equivalent
20. footer 50x30mm angle iron, metalized and painted
21. shot-profile cove like pvc type "bolta" or equivalent
22. ceiling boards - gypsum board "knauf" or equivalent
23. thermal insulation in expanded polyurethane
24. handmade brick facings torn in half, "s.brás de alportel" or equivalent, together with 2cm, including the execution of mortar and plaster treatment of surfaces
25. brickwork 11cm
26. structure slatted iron tubes R4S 30x30 metallized
27. concrete block (structural project)
28. zinc sheet-system stitched together
29. membrane "delta vm zinc" or equivalent
30. plywood 12mm
31. sill and lintel in sheet metal 2mm metallic and painted
32. sill and lintel in sheet metal 2mm metallic and painted on slatted structure in iron tubes R4S 30x30 metallized

详图1 detail 1

详图2 detail 2

1. rufus and security clamps in zinc
2. thermal insulation plates in rigid extruded polystyrene foam with e=4cm, "wallmate" or equivalent
3. concrete beam
4. brickwork 3cm
5. handmade brick facings, torn in half, "s.brás de alportel" or equivalent, together with 2cm, including the execution of mortar and plaster treatment of surfaces
6. plaster "seral" or equivalent
7. brickwork 15cm
8. brickwork 11cm
9. ceiling boards - gypsum board "knauf" or equivalent
10. plaster "seral" or equivalent, to paint
11. lintel sheet metal with 1cm, metallized and painted
12. double glazed sash type "maciça" or equivalent for enamelling
13. system self-supporting glass panels comprising glass u-shaped section such as "profilit u-glass" or equivalent
14. floor "sikafloor 261 - autoalisante" or equivalent

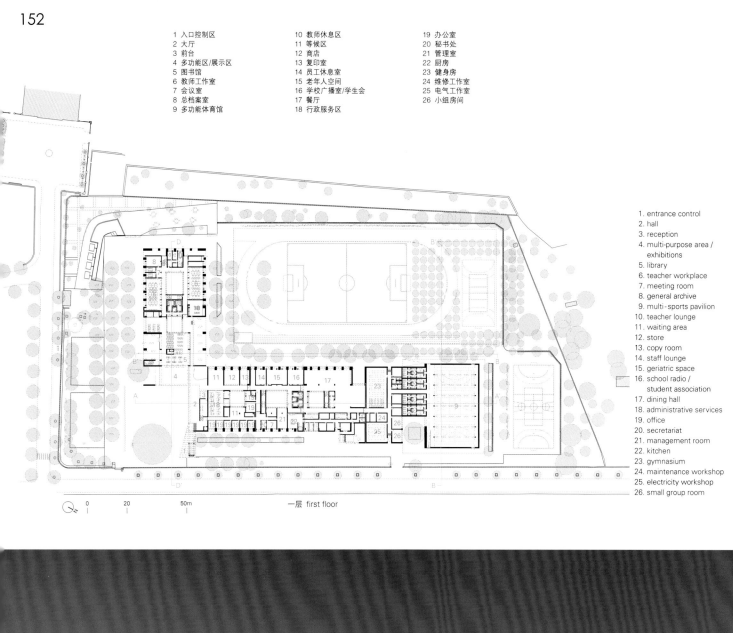

1 入口控制区	10 教师休息区	19 办公室
2 大厅	11 等候区	20 秘书处
3 前台	12 商店	21 管理室
4 多功能区/展示区	13 复印室	22 厨房
5 图书馆	14 员工休息室	23 健身房
6 教师工作室	15 老年人空间	24 维修工作室
7 会议室	16 学校广播室/学生会	25 电气工作室
8 总档案室	17 餐厅	26 小组房间
9 多功能体育馆	18 行政服务区	

1. entrance control
2. hall
3. reception
4. multi-purpose area / exhibitions
5. library
6. teacher workplace
7. meeting room
8. general archive
9. multi-sports pavilion
10. teacher lounge
11. waiting area
12. store
13. copy room
14. staff lounge
15. geriatric space
16. school radio / student association
17. dining hall
18. administrative services
19. office
20. secretariat
21. management room
22. kitchen
23. gymnasium
24. maintenance workshop
25. electricity workshop
26. small group room

一层 first floor

1 储藏室	6 学生卫生间
2 教室	7 物理实验室
3 生物/化学实验室	8 小组房间
4 制备室	9 视觉/技术教学室
5 多功能实验室	

1. storage
2. classroom
3. biology / chemistry laboratory
4. preparation room
5. multipurpose laboratory
6. student toilets
7. physics laboratory
8. small group room
9. visual/technological education room

三层 third floor

1 戏剧室	9 学生卫生间
2 多功能房间	10 技术区
3 公用电话亭	11 视觉/技术教学室
4 教室	12 美术室
5 音乐室	13 多媒体/设计室
6 小组房间	14 科学实验室
7 信息通信技术室	15 造型艺术工作室
8 储藏室	16 几何室

1. theatre room
2. multipurpose room
3. booth
4. classroom
5. music room
6. small group room
7. ICT room
8. storage room
9. student toilets
10. technical area
11. visual / technological education room
12. drawing room
13. multimedia/design room
14. science laboratory
15. plastic arts workshop
16. geometry room

二层 second floor

Vila Viçosa High School

The intervention area consisted of a school ground, vast and arid, where, between the existing buildings, three large pine trees grew. The central block had social and management functions; other two held classrooms and laboratories, another was still devoted to workshops and, finally, there was a sports pavilion. Functionally outdated, formally unattractive and structurally fragile, it denoted a prematurely aging equipment and an obvious failure to be warm or friendly enough for the school community. Beyond the aspects listed above, it also became urgent and essential to solve the thermal, acoustic and air handling problems, as well as architectural barriers, disorder of outdoor spaces, etc.

This reality is the starting point for a new school concept based on proposal. Firstly, the buildings to preserve were selected, and the drawing found its way into a unique final object. Maintaining both classroom buildings and the sports pavilion, the solution became to fulfill the spaces between them, creating new volumes and thus becoming a continuous element.

This "new" school building re-organized, was started by marking its entrance and a dignified hall. That space extended itself outdoors, joining an exterior square. On the inside, a simple and bright image was required, with obvious paths, easing the connection between the several operational entities.

Externally, a solid artisan brick covering provided protection of the building, reducing maintenance costs, while giving unity to the whole complex. The Glass Reinforced Concrete (GRC) grid allowed the control of light intensity and provided uniformity to the outside image of the building, boosting the integration of pre-existing blocks. The entire surrounding area was redesigned, enhancing outdoor activities, the relationship with the interior spaces and assuming an improved framework for the building.

Cândido Chuva Gomes Arquitectos

项目名称：Públia Hortênsia de Castro High School
地点：vila viçosa, évora, portugal
建筑师：Cândido Chuva Gomes
项目团队：Sandra Amorim, Ana Florêncio, Ana Luísa Brandão, Bruno Bernardo, Bruno Ribeiro, Cristina Martinho, Paulo Melo, Vanda Vitorino, Luís Quaresma, Lénia Nobre, João Seabra
项目管理：Cândido Chuva Gomes Arquitectos, Lda. / Ana Pestana, Unipessoal Lda.
建筑承包商：Lena Construções / Abrantina / MRG Engenharia e Construção
设备工程师：Joaquim de Almeida
电气工程师：Hélder Reis
照明设计师：Rogério Oliveira
景观设计师：Carlos Correia Dias, Elsa Barroso, Elsa Barroso, Armando Conceição, Lúcia Barreto, Cristina Matias
承包商：ena construções / abrantina / mrg engenharia e construção
甲方：Parque Escolar-EPE
用地面积：31,300.58m² 建筑面积：14,570.26m² 总楼面积：11,771.60m²
施工时间：2008—2011 竣工时间：2011
摄影师：courtesy of the architect – p.146~147, p.152~153
©José Campos Moreira – p.148, p.150, p.151, p.154~155

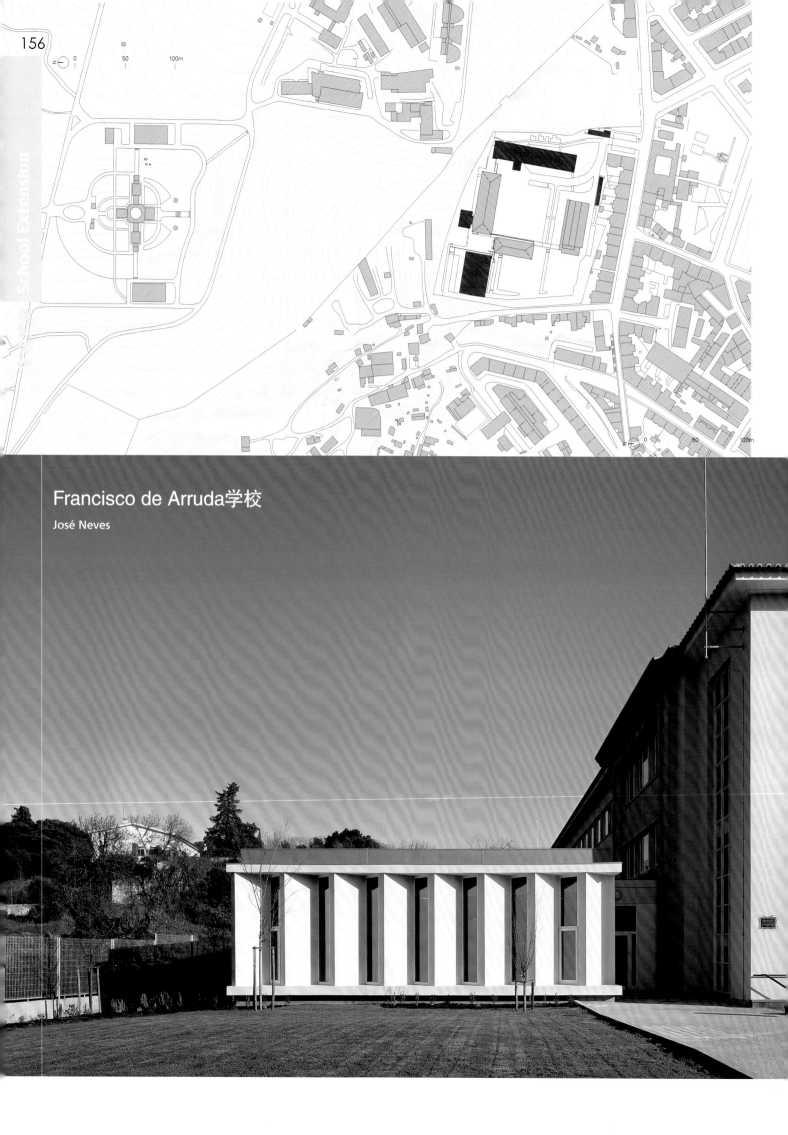

Francisco de Arruda学校

José Neves

该项目包括对三栋原有建筑的修复和三种新元素的糅合。这所学校最初由建筑师约瑟·安东尼奥·佩德罗索于1956年设计。学校的三座建筑坐落在山坡上,与农学院的田园相连接,可远眺阿尔坎塔拉地区。

原始建筑占据了山坡上的三个平台,围绕着一个中央天井排布,该建筑由三个部分组成,每个部分都有自己的功能——教学、行政以及带顶的操场;体育馆和自助餐厅;工作室——这三个部分由带顶的室外走廊连接。建筑这三个组成部分具有完全不同的外形,建筑师记录着它们的成长,仿佛它们是在历史中逐步建成的,形成一座小城市。建筑师的工程默默地延续着这种历程。

该项目的室外区域非常宽阔,通过新种植园来突出场地内的许多植被的连续性,以强调建筑场地成为公共花园的感觉。建筑师设计的楼梯打开了原有的死巷,将中央天井和位于顶部平台的Caravelas旧天井连接起来。新图书馆和横跨带顶操场的运动场地之间的直接连接与上述连接都是为了强调这个项目作为一个公共花园的整体感和深刻塑造学校的日常生活而进行的改造。另一方面,通过嵌入最小的结构,特别是外观和室内空间的饰面,原始建筑既得到了修复,又可以适应新的需求。

最终,该项目增加了三种新元素:

a) 一座长长的两层建筑完善了原始天井的轮廓。穿过公园的一条小径就能看到该建筑,同时它也是新学校的入口。该建筑被设计为公共画廊、会议室和学习区,沿着庭院设置,就好像一条封闭的拱廊,建筑内部还包括一层的新图书馆和顶层的新实验室,东侧面向学校的花园、周边田园和城市;

b) 淋浴间建在Caravelas旧天井中,Caravelas旧天井位于原有体育馆和新建带顶的运动场之间的下部,其中运动场坐落在场地的一个最高平台上;

c) 从未被作为中庭使用的原始主中庭,目前是一个小型多功能房间的中庭,它被设计成公园亭阁,学校内的人们将其称为"魔法立方"。

在赋予Francisco de Arruda学校旧址以美感后,João dos Santos明确而直接的话语再一次对工程起到了更大的激励作用:"教育首要的是其所在的自然而审美协调的环境。"

Francisco de Arruda School

The project consists of the rehabilitation of three existing buildings and the integration of three new elements. This school was originally designed by architect José António Pedroso in 1956. The three bodies that constituted the school are situated on the hillside in the continuity of the Faculty of Agronomy fields, overlooking the district of Alcântara.

The original building, spreading across three platforms modeled on the hill, was organized around a central patio and had three bodies each with its own functions – teaching, administration, and covered playground; gym and cafeteria; workshops – linked through covered outdoor galleries. These bodies, with very different shapes, were taken by architects as if they had been built over time, making a little city. Their project quietly continues this process.

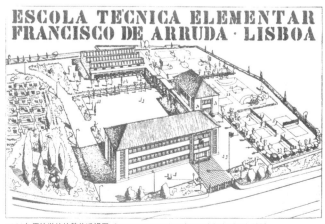

1956年原始学校的整体透视图 General perspective of the original school, 1956

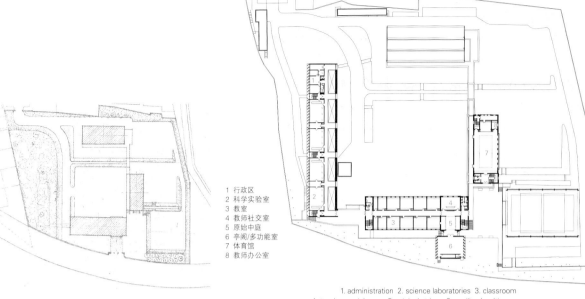

1 行政区
2 科学实验室
3 教室
4 教师社交室
5 原始中庭
6 亭阁/多功能室
7 体育馆
8 教师办公室

1. administration 2. science laboratories 3. classroom
4. teacher social room 5. original atrium 6. pavilion/multipurpose room
7. gym 8. teacher workroom

三层 third floor

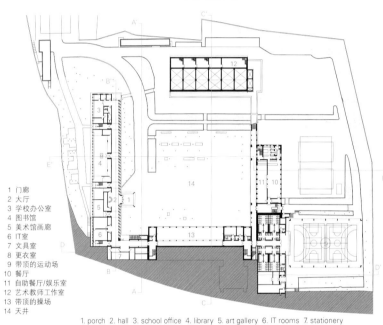

1 门廊
2 大厅
3 学校办公室
4 图书馆
5 美术馆画廊
6 IT室
7 文具室
8 更衣室
9 带顶的运动场
10 餐厅
11 自助餐厅/娱乐室
12 艺术教师工作室
13 带顶的操场
14 天井

1. porch 2. hall 3. school office 4. library 5. art gallery 6. IT rooms 7. stationery
8. changing rooms 9. covered sports field 10. refectory 11. cafeteria/recreation
room 12. art teacher workrooms 13. covered playground 14. patio

二层 second floor

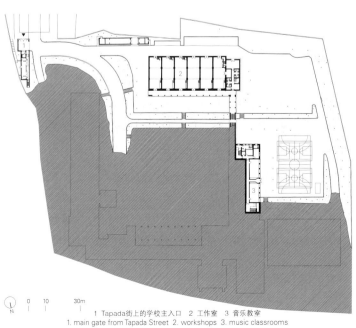

1 Tapada街上的学校主入口 2 工作室 3 音乐教室
1. main gate from Tapada Street 2. workshops 3. music classrooms

一层 first floor

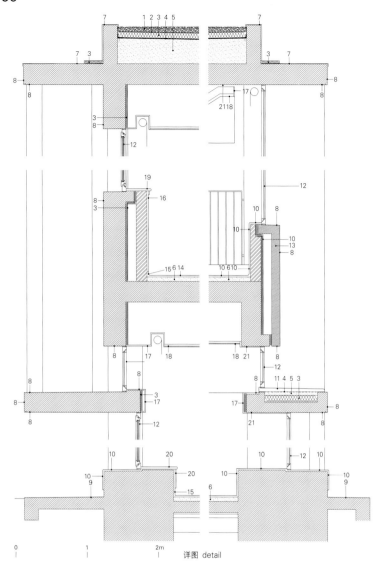

1. layer of gravel
2. geotextile membrane
3. extruded polystyrene foam
4. waterproofing pvc panel
5. lightweight concrete
6. reinforced mortar bed
7. sheet-zinc covering
8. cement based smooth plaster with plastic paint finish
9. sealed concrete
10. marble agglomerate
11. lioz limestone
12. wood window frame with synthetic paint finish
13. precast concrete element
14. epoxy-resin floor coating
15. epoxy coating
16. cement based smooth plaster with acrylic paint finish
17. gypsum board with plastic paint finish
18. suspended cement-bonded wood fiber acoustic panel with plastic paint finish
19. agba solid wood windowsill with synthetic paint finish
20. plywood with synthetic paint finish
21. gypsum based plaster with plastic paint finish

详图 detail

For the outdoor areas, of unusually generous dimensions, the project underlines the continuity of the vegetation of the fields through new plantations, in order to emphasize the sense of the site as a public garden. The staircase that they designed, opening an existing cul-de-sac to connect the central patio to the old Patio das Caravelas situated on the top platform, and the direct connection created between the new library and the sports fields across the covered playground, are both transformations that help to emphasize the sense of the whole as a public garden and profoundly shape the daily living in the school. On the other hand, the original buildings are restored as well as adapted to new needs, through minimum interventions, particularly in the external form and the finishes of the interior spaces.

Finally, three new elements are added:

a) A long two-story building completes the outline of the original patio. It is discovered after a short path through the garden, as the new school entrance. Designed as a public gallery, meeting and study space, along the courtyard – as a sheltered stoa – this building contains the new library on the ground floor and the new laboratories on the top floor, facing, eastward, the gardens of the school, the neighboring fields and the city;

b) The shower spaces are built under the old Patio das Caravelas between the existing gym and the new covered sports field that is situated in one of the high platforms of the site;

c) The original main atrium, never used as one, is now the atrium of a small multipurpose room, designed as a garden pavilion, which the population of the school calls "magic cube".

Given the beauty of the original Francisco de Arruda School situation, once again the clear and direct words of João dos Santos were served as a greater stimulus to the project: "What is important in education, first of all, is the natural and aesthetically harmonious environment in which it takes place." José Neves

项目名称：Francisco de Arruda School
地点：Lisbon, Portugal
建筑师：José Neves
项目团队：Rui Sousa Pinto, Ana Belo, André Matos, Filipe Cameira, Martim Enes Dias, Nuno Florêncio, Steven Evans, Vitor Quaresma; João Pernão, Maria Capelo(color consultants)
结构工程师：BETAR
电气工程师：Silvino Maio & Lacerda Moreira
消防安全顾问：António Portugal
景观建筑师：F|C Arquitectura Paisagista
甲方：Parque Escolar-EPE
用地面积：25,426m²
修建结构面积：8,101m²
新建结构面积：5,174m²
设计时间：2008—2009
施工时间：2009—2011
摄影师：©Laura Castro Caldas & Paulo Cintra (courtesy of the architect)
- p.157 top, p.158~159, p.160, p.162, p.163
©João Morgado (courtesy of the architect) - p.156~157, p.161

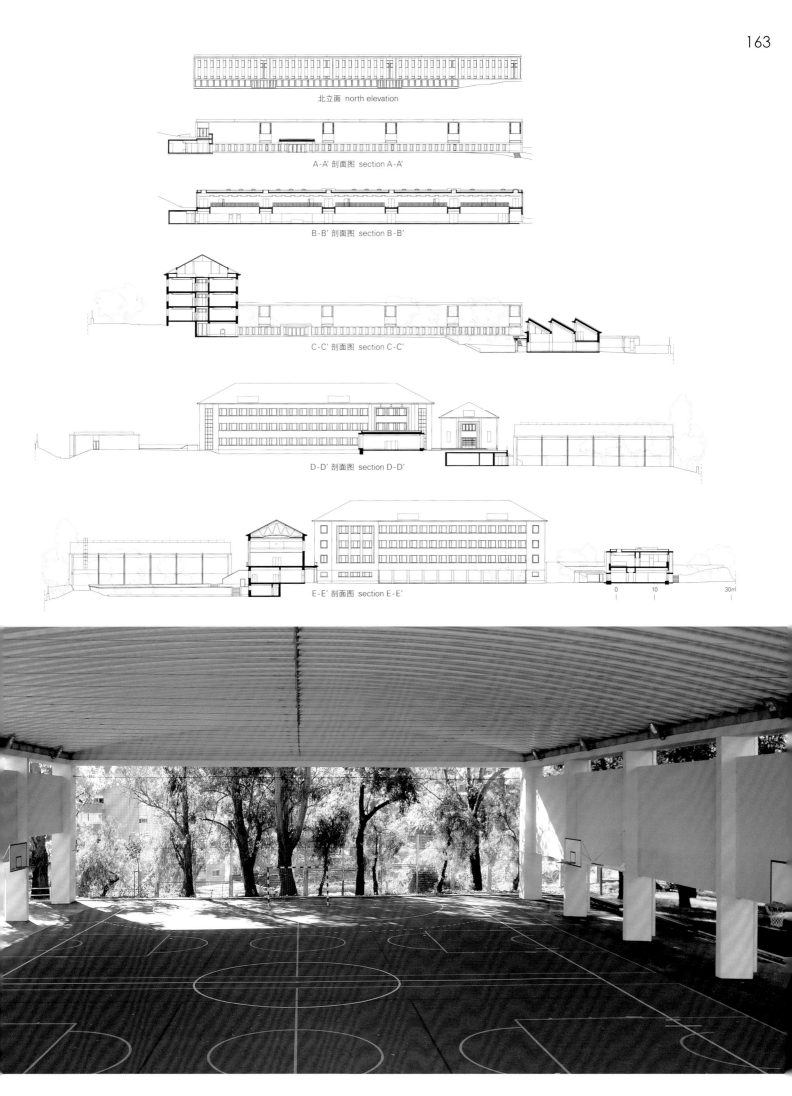

拉滕贝格中学

Architekt Daniel Fügenschuh ZT GmbH

拉滕贝格是奥地利最小的城镇,位于提洛尔的低洼地带,人口仅有440人。该城镇努力保存着目前几乎完好无损的中世纪后期风景如画的城市氛围。对这所中学的扩建是必要的,因为学校建筑太小,无法满足现在午间看护学生的需求。这只是一种试图与周边环境相融合的建筑类型。

新建筑非常现代化。同时,它还灵巧地嵌入了历史建筑群中,通过应用古老的比例、颜色、形状和材料,呈现出完美的效果。

学校新建筑的场地非常敏感。

建筑位于原奥古斯丁修道院和1970年扩建的体育馆之间。哥特式修道院于17世纪经历了宏伟的巴洛克式改建。从20世纪70年代开始,修道院的部分建筑转变成为宗教艺术博物馆和中学。在其较短的一侧,连接着学校和旧城镇的细长的新建筑通过一座小型玻璃桥,以一个恰当的角度与修道院相连,同时在右侧,建筑横亘在带有易碎的玻璃屋顶的体育馆上方。新建的局部带顶的户外空间和学校庭院通过抬升建筑打造而成。庭院的铺石路非常有城市感。它有很多种潜在的用途,例如是举办半公开活动的理想场地。

夸张的洞口构成了拉滕贝格新学校的独特魅力。沿街的狭窄立面上有一面巨大的方形窗,窗户是雕刻的,用大量的铜包裹,与墙体齐平,如同一种具有代表性的陈列窗。

在较长的立面上,窗户与正立面的陈列窗布局方式不同,它们尺寸很小,看上去如同黑洞,与室内混凝土墙面齐平设置。

体育馆自身包含了很多重要的嵌入元素。翻新后,该空间为学校活动提供了一处吸引人的场地。

体育馆不仅进行更新,以满足最新的建筑标准,其基础设施也满足举办时下学校体育活动的要求,而且它的两个侧立面可打开——一侧通向狭窄的走廊,另一侧通向保留下来的花园,该花园曾经是美丽的旧式修道院花园,周边采用石墙围护。

明亮的色彩、玻璃和木材在室内发挥了重要的作用。教室的墙体采用了不常见的清水混凝土,同时以一种不常见的方式在木框架中进行浇筑,从而在混凝土表面永久地印上了明显的木质纹理。这给学校带来了令人愉悦的视觉效果,但也带来了一定的副作用——这些墙体又将声音反射到吸音天花板上。

Rattenberg Secondary School

Rattenberg, Austria's smallest town, is located in the Tyrolean lowland with a population of 440 people. It has managed to preserve its picturesque late-medieval ambience to the present day mostly undamaged. The extension to a secondary school was needed as the building became too small to meet today's demands of afternoon care for the pupils. It is nothing but the kind of architecture that attempts to integrate itself in its environment.

It is decisively contemporary. At the same time, the new building is sensitively inserted into the historical ensemble which reflects in a wonderful way thanks to a playful use of historic proportions, colors, shapes and materials.

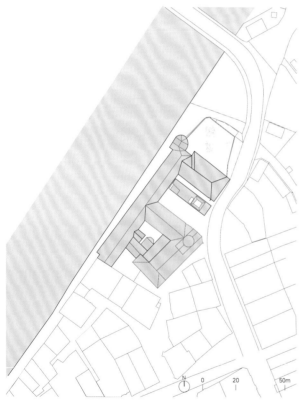

玻璃走廊详图
glass corridor detail

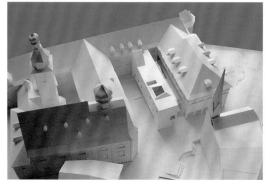

西南立面 south-west elevation

A-A' 剖面图 section A-A'

B-B' 剖面图 section B-B'

项目名称：Hauptschule Rattenberg
地点：Rattenberg, Österreich
建筑师：Daniel Fügenschuh
项目团队：Thomas Niederberger, Isabella Dorigo
机械工程师：TAP
结构工程师：INGENA
甲方：Rattenberger Immobilien GmbH
用地面积：1,492.45m²
建筑面积：250m²
有效楼面面积：1,807.3m²
设计时间：2009
竣工时间：2011
摄影师：©Christian Flatscher (courtesy of the architect)

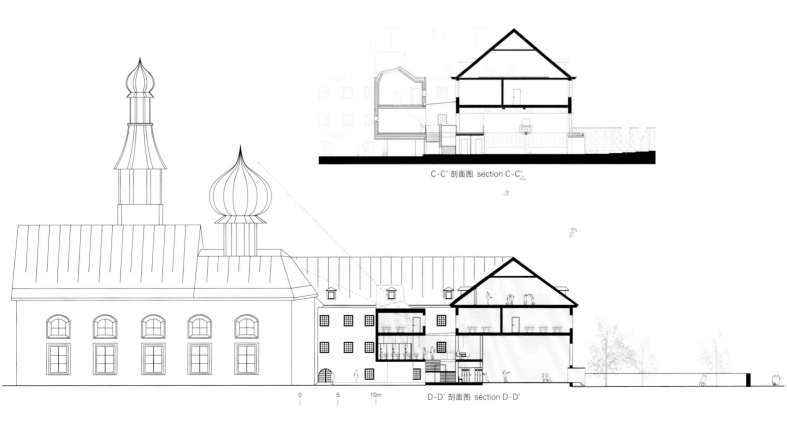

C-C' 剖面图 section C-C'

D-D' 剖面图 section D-D'

The site of the new school building is a sensitive one.
It is located between a former Augustine monastery and a gym which was extended in 1970. The Gothic monastery underwent a magnificent Baroque transformation in the 17th century. Since the 1970s it has been transformed partly into a museum of religious art and a secondary school. On its short side the slender new building which connects school and old town is linked to the monastery at a right angel through a small glass bridge, while to the right it docks onto the gym with a fragile glass roof construction. A new, partly covered outdoor space and school yard are gained through raising the building from the ground. Its paving gives a highly urban impression. It suggests a variety of possible uses, and is ideal for semi-public events, for instance.

Extravagant openings constitute the particular charm of the new school in Rattenberg. In the narrow facade onto the street a single large, square window, opulently surrounded in copper, is incised flush with the wall, as a kind of representative display window.

On the long facades the windows are set differently to the display window in the front facade. They are small and appear like dark holes as they are fitted flush to the inner concrete face.

A number of important interventions were made in the gym itself. After its refurbishment the space offers an attractive location for school events.

The gym was not only updated to meet the newest building standards with its infrastructure adapted to meet the current demands of school sporting activities, but both of its side facades were opened up – on one side to a narrow gallery, on the other side to the remaining garden once a lovely old monastery garden surrounded by a stone walls.

Bright colors, glass and wood play an important role in the interior. As does fair-faced concrete of an unfamiliar kind: the walls of the classrooms were poured in a wooden formwork so that the striking wood grain is immortalized in the surface of the concrete. A delightful visual effect is created with the side-effect of particular relevance for the school – these walls reflect sound to the sound absorbing ceiling. Architekt Daniel Fügenschuh ZT GmbH

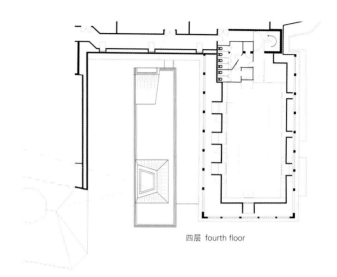

四层 fourth floor

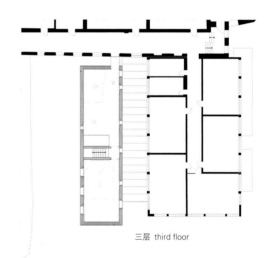

三层 third floor

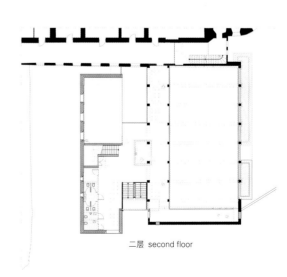

二层 second floor

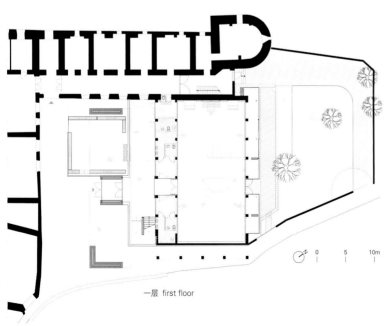

一层 first floor

贝尔纳多特学校的扩建
Tegnestuen Vandkunsten

范德昆斯坦为哥本哈根北侧的小型贝尔纳多特学校完成了占地235m²的扩建工程。

贝尔纳多特学校于1949年由一群有着共同理念的父母所创办：通过促进国际间的了解，来避免未来矛盾的产生。

从教育方面来说，这所学校从一开始便有着十分明确的目标：希望打破传统学校教育的刻板与专制，实行以孩子为中心的新教育理念。虽然学校自创办以来针对现代社会的不同需求不断发展，且适应了大量状况，但是其指导方针仍紧紧围绕着创造一个特殊的学习环境而定。

学校位于过去数年里逐渐增建的几座旧别墅的中间，创造了一个充满特色的迷宫式教学环境，但却缺乏设施和基本的室外空间。

范德昆斯坦的新建筑增加了一座图书馆和一个带舞台设施的体育馆，还扩大了原有的教室和工作室，同时新建筑的屋顶露台为孩子们提供了更大、更集中的活动场地。所有新空间都与原来的入口和交通流线区相连接。

该建筑是采用木材和钢材的轻质结构，保温性能很高。项目采用了原有建筑的迷宫特征作为场地的一个基本方面，而范德昆斯坦的品牌黑色覆层的运用使其获得了某种视觉上的连贯性，覆层上的不锈钢丝与沿着新房屋种植的三种爬墙植被相映成趣。

随着时间的流逝，建筑立面将变成绿色，为植被后面的室内房间过滤阳光。

Bernadotte School Extension

Vandkunsten has completed a 235m² extension to the small Bernadotte School in the north of Copenhagen.
Bernadotte School was opened in 1949 by a group of parents who were motivated by the same idea: the prevention of future conflict by the promotion of international understanding.

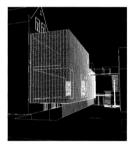

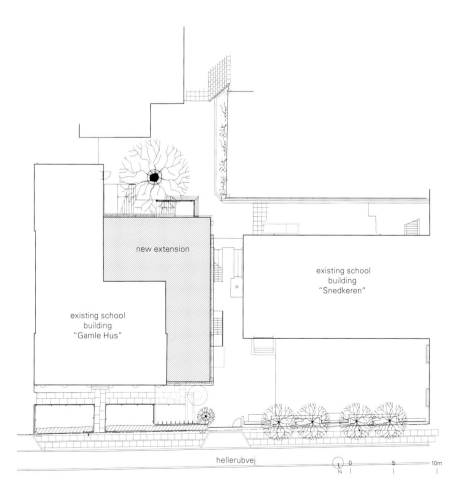

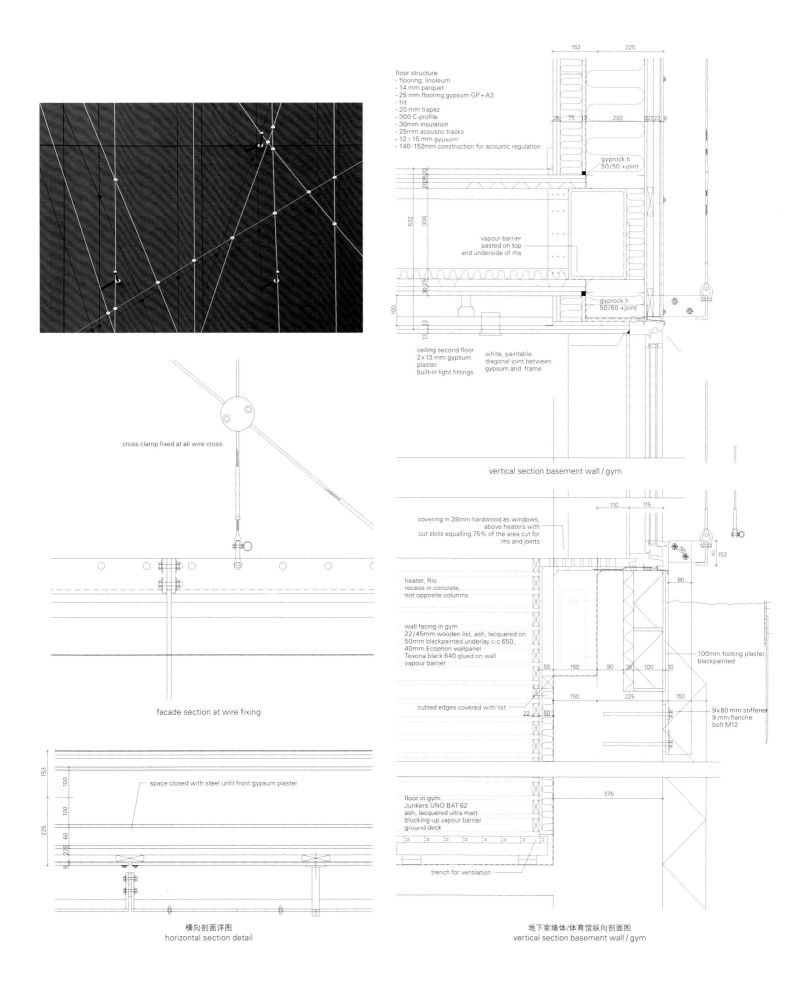

Educationally, the school had very clear aims from the beginning: a desire to break away from the rigidity and authoritarianism of traditional schooling, and to put the new ideas of child-centred pedagogy into practice. Though the school has developed and adapted a great deal since those days in response to the varying demands of modern society, the guiding principles have held fast and still help to create a very special learning environment.

The school is housed in a couple of old villas that have been added over the years, creating a labyrinthine teaching environment full of characters but lacking in facilities and basic outdoor spaces. Vandkunsten's new building adds a library and a gym with stage facilities, and also extends existing classrooms and workshops while new roof terraces leave a larger and more articulated play area for the children. All new spaces tie into existing access and circulation.

The construction is lightweight in wood and steel, and insulation values are high. While the project accepts the maze-like quality of the building as a fundamental aspect of the place, a degree of visual coherence has been achieved with Vandkunsten's trademark black cladding – and with the stainless wires in place for the three varieties of creepers that have been planted along the new house. As time passes, the facade will turn into green and provide the rooms behind with filtered daylight. Tegnestuen Vandkunsten

项目名称：Bernadotte School Extention
地点：Copenhagen, Denmark
建筑师：Tegnestuen Vandkunsten A/S
项目团队：Olmo Ahlmann, Jens Thomas Arnfred
工程师：Moe & Brødsgaard A/S
承包商：DrivhusEffekten ApS
甲方：Bernadotteskolen
用地面积：235m²
竣工时间：2009
摄影师：©Adam Mørk (courtesy of the architect)

西立面 west elevation

北立面 north elevation

南立面 south elevation

A-A' 剖面图 section A-A'

B-B' 剖面图 section B-B'

1 屋顶露台
2 教室
1. roof terrace
2. classroom

二层 second floor

1 屋顶露台
2 教室
1. roof terrace
2. classroom

四层 fourth floor

1 体育馆及舞台
2 技术室
3 工作室
1. gym and stage
2. technique
3. workshop

一层 first floor

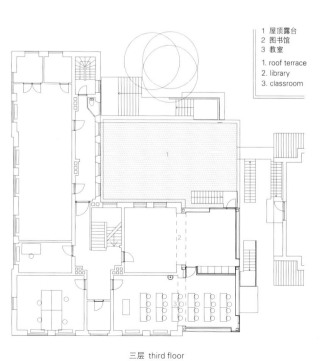

1 屋顶露台
2 图书馆
3 教室
1. roof terrace
2. library
3. classroom

三层 third floor

裴斯塔洛齐学校

SOMAA + Gabriele Dongus

新裴斯塔洛齐学校坐落于德国斯图加特市周边的一座名为莱昂贝格的小城中。

莱昂贝格是以建筑师弗雷·奥托的故乡而闻名,而这所新学校却与奥托著名的轻质结构形成了对比。该学校局部采用厚重的混凝土墙,建筑可俯瞰溪谷,坐落在普通的独栋住宅单元、60年代的社区中心和约1900年建造的砖结构的裴斯塔洛齐学校之间。

裴斯塔洛齐学校是一所为学习有障碍的孩子设立的小学,被设计成一种可以采用不同方式进行学习的环境。通过利用场地的地形,建筑师创造出开放式和封闭式空间,保护孩子免受外部干扰,促进校园内的玩耍和社交互动。建筑建在场地的一片斜坡上,它与街道的相邻处设有封闭的板状粗糙混凝土立面。在建筑的另一面,封闭的临街立面与面对着老裴斯塔洛齐学校建筑的开放式玻璃立面相对应。建筑还形成了一个庭院,在新旧学校之间建立了可视联系。

在室内,建筑被分成三个部分:交通流线区、服务功能区和教室。

交通流线区由平缓蜿蜒的淡紫色走廊组成,走廊跨越了建筑临街部分的整个长度,引导人们走向教室和服务区。走廊非常宽,还带有休息和娱乐的辅助区域。圆形的大型天窗为走廊空间提供了自然光。淡紫色和地板上与其互补的绿色营造了一种亲切而开放的氛围。走廊尽头的场地上划出了一间私人户外教室,在建筑内形成了一种带有保护性的开放感觉。

服务空间位于走廊和教室之间,有利于为每间教室形成宽敞的入口空间。这些空间的表面带有浓重的绿色,也使得每间教室入口都不尽相同。每间教室都沿着曲形玻璃立面排布。人们还能在教室中看见旧学校和溪谷中的独特风景。通高的窗户以及额外的圆形天窗都提供了足够的自然光,所以上课时间很少需要人工照明。所有的教室也都设有直接进入新操场的入口,新操场的长度跨越了整个玻璃立面。与户外和充足自然光线的直接联系可以鼓励学生们学习,提高他们的专注力。

在建筑临街的一侧，建筑师通过改进板状混凝土的传统浇筑方法来建造封闭式的混凝土立面。该结构由两块混凝土板组成，混凝土板由保温材料隔开。立面框架是由三种不同尺寸的木板相叠成形的。这个图形内嵌入了特殊的木板，用来旋转后面的木板。这使建筑看起来好像斜靠着地形一样。框架木板为粗齿锯木，产生了有着粗糙纹理的精细而锋利的波状立面。

将场地所带来的挑战转化为其优势，建筑师可以在需要隐私的地方将空间封闭，在需要社交互动的地方将空间开放。最终形成的设计是一座创新型学校，它创造了一种位于新旧学校之间的社区的感觉。

Pestalozzi School

The new Pestalozzi School is located outside of Stuttgart, in the small city of Leonberg, Germany.

While Leonberg is known for being the home of Architect Frei Otto, the new school is a contrast to the lightweight structures for which Otto is famous for. The school, built partially of massive concrete walls, overlooks a valley and sits in the midst of ordinary, single-family housing units, a community centre from the sixties and the brick Pestalozzi School, constructed around 1900. The Pestalozzi School, an elementary school for children with learning disabilities was designed to be an environment conducive to a different way of learning. By using the topography of the site, the architects created open and closed spaces that protect the children from the exterior and promote play and social interaction within. The building is set into the slope of the site and has a closed, rough and board-formed concrete facade bordering the street. On the opposite side of the building, the closed street facade contrasts with an open glass facade facing

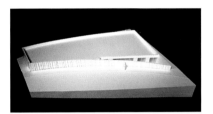
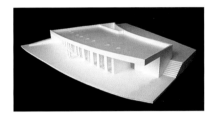

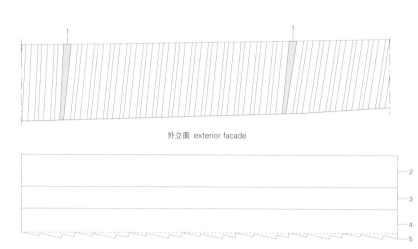

外立面 exterior facade

外立面的水平剖面详图
horizontal section detail of exterior facade

1. points at which the facade is skewed 3 degrees
2. inner concrete shell 20cm
3. insulation 14cm
4. outer concrete shell 15 + 3.7
5. board-formed concrete
6. formwork
7. board-formwork 1-2.4 x 1.9
8. board-formwork 2-3 x 16.4
9. board-formwork 3-1 x 20
10. at different points on the facade, an alternative board is inserted and causes the subsequent formwork to rotate 3 degrees

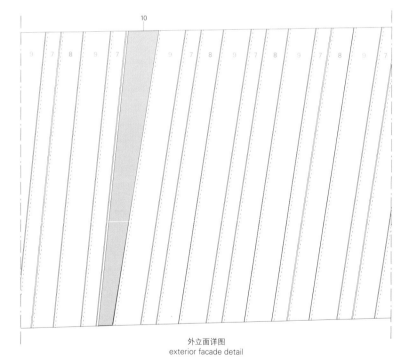

外立面详图
exterior facade detail

the old Pestalozzi School building. The building creates a courtyard and establishes a visual relationship between the old and new school.

On the interior, the building is divided into three parts: the circulation, the service functions, and the classrooms.

The circulation consists of a gently curving lilac corridor that spans the length of the street side of the building and connects occupants to the classrooms and service spaces. It has generous width with an ancillary space for resting and playing. Large circular skylights give the space natural light. The lilac color and complementary green of the flooring contribute to a friendly and open atmosphere. At the end of the corridor, a private outdoor classroom has been carved out of the site and creates a feeling of protected openness within the building.

The service spaces, located between the corridor and the classrooms help to generate generous entrance space for each classroom. Surfaced in intensifying shades of green, these spaces also differentiate the entrances to the classrooms. Each classroom is located along the curving glass facade. It has a prominent view of the old school and the valley. The floor-to-ceiling windows and additional circular skylights provide generous natural light, so artificial lighting is rarely needed during class time. All classrooms also have direct access to the new playground, which spans the length of the glass facade. The direct connection with the outdoors and abundance of natural light encourages learning and concentration among students.

一层 first floor

A-A' 剖面图 section A-A'

On the side of the building facing the street, a closed, concrete facade was created by altering the traditional method of board-formed concrete. The construction consists of two shells of concrete separated by insulation. For the facade formwork, three different board sizes were overlapped to create a pattern. Within this pattern, special boards were inserted and used to rotate the subsequent boards. This causes the building to look as if it is leaning against the topography. The boards used for the formwork were rough-saw and the result is a subtle, yet sharp and undulating facade with a rough texture.

By using the challenging aspects of the site as their advantage, the architects were able to close spaces where privacy was needed, and to open spaces where social interaction was welcomed. The result is an innovative school design that creates a feeling of community between the new and old school.

SOMAA + Gabriele Dongus

项目名称：Pestalozzi School
地点：Leonberg, Germany
建筑师：Tobias Bochmann, Hadi A. Tandawardaja, Gabriele Dongus, Serpil Erden
结构工程师：Ingenieurburo Dieter Urbanski, Leonberg
建筑物理：Bauphysik 5, Backnang
电气工程师：PEN Planung Engineering Nick GmbH, Leonberg
木匠：Projektholz, Kirchheim unter Teck
油漆工：Anton Geiselhart GmbH & Co.KG, Reutlingen-Pfullingen
结构施工：Wildermuth, Bietigheim-Bissingen
金属结构：Trappe Metallbau, Bad Liebenzell
立面：Schreinerei Otto Obermüller, Winnenden
学校家具：VS Vereinigte Spezialmöbelfabriken GmbH & Co. KG, Tauberbischofsheim
甲方：City of Leonberg, Building Management, Christian Beutelspacher
用地面积：680.5m² 总建筑面积：479.4m²
有效楼面面积：428.4m² 总体积：1,610m³
竣工时间：2010
摄影师：©Patricia Neligan (courtesy of the architect)

>>20
Future Systems+Shiro Studio

Jan Kaplický[left] was born and raised in Prague. Worked for Renzo Piano and Richard Rogers. In 1979, he founded the architectural practice Future Systems with David Nixon and he began to develop an architectural style that will combine organic shapes with a high-tech futurism. Suddenly passed away in Prague in 2009, where he went as the winner of the design competition for Prague's new National Library.
Andrea Morgante[left] was born in Milan. After working in RMJM London he joins Future Systems architects in 2001 where in 2006 he becomes Associate Director for almost 8 years. Works with Jan Kaplický, founder of Future Systems. In 2009 establishes Shiro Studio in London. Parallel to Architecture Andrea collaborates with design companies like D_Shape, DND, Alessi, Agape Design, MGX, Poltrona Frau. Also regularly teaches and lectures across Europe.

>>100
Antonio Ravalli Architetti

Antonio Ravalli graduated from University of Firenze, Faculty of architecture in 1988. Started professional career in "Ravalli Architetti Associati" studio and collaborated with "Studio Aleardi" from 1989 to 1992. Has been teaching architectural design in Ferrara University's faculty of Architecture since 1994.
Organizes and manages workshops on urban requalification and is often invited to hold conferences and lectures in many European and American cities.
Antonio Ravalli is principally interested in the transformation process of the urban and territorial landscape, both at a macro and a micro scale, where architecture has the possibility to investigate alternative solutions to the multiple conditions of reality.

>>48
Langarita-Navarro Arquitectos

María Langarita[right] graduated from Universidad de Navarra. Has been an associate lecturer at the Architectural Projects Departament at the Universidad de Navarra and at the ETSA Madrid.
Víctor Navarro[left] graduated from the ETSA Madrid. Works as lecturer in Architectural Projects at the Universidad Europea de Madrid. Is co-founder with Roberto González of Urgente, defined as periodic publication for the dissemination of documents that may be of current interest.
María Langarita and Víctor Navarro have been working together since 2005. Won several competitions and have been invited to numerous conferences and workshops. Have also been invited as the jury in the Santo Domingo Bienale.

>>92
Bang Architectes

Was founded by Nicolas Gaudard and Nicolas Hugoo. Nicolas Gaudard graduated from the Paris Belleville Architecture School and Nicolas Hugoo graduated from Lille Architect School. Their work is based on the belief on a new kind of architecture; innovative, attractive and intelligent, inspired by the needs of the new era, getting over the great dogmas of the twentieth century.

>>82
**Andrés Holguín Torres
+ David R. Morales Hernández**

Is based in Venice since 2006. Andrés Holguín Torres was graduated from the Architecture University of Venice (IUAV) and the Universidad de los Andes (Bogotá, Colombia). Was granted a master degree in Urban Project at the Universitat Politécnica di Catalunya (Barcellona, Spain). Ph.D. in Urban planning at the the University of Venice (IUAV). Andrés Holguín Torres is specialized in urban and architectonic projects as well as public spaces.
David R. Morales Hernández graduated from the Universidad de los Andes in Bogotá in 1998. Office based in Barcelona since 2004, and recently opened a new office in Colombia. Works on the development and rehabilitation of housing and middle size projects.

>>130
BFJ Arquitectos
Was founded in 1998 by its partners, Francisco Amaral Pólvora, Bernardo Campos Pereira and José Amaral Pólvora. Won Leca Building Awards in 2004 and the Eugénio dos Santos Architecture Award in 2007 for the Cine-Teatro de Alcobaca.
Francisso and José studied at the Technical University of Lisbon, Portugal. Bernardo studied at the University of Waterloo School of Architecture in Canada. Each of them has collaborated with other architects independently before they set up their own firm.

>>60
Ector Hoogstad Architecten
Joost Ector is a partner and design principal of this company. Graduated cum laude in 1996 from Eindhoven University of Technology and then joined Hoogstad Architecten. Became project architect in 1999 with his winning design for the HES Amsterdam, and then a director of the firm in 2001 and in the following year Joost Ector became co-owner with Max Pape of what is now Ector Hoogstad Architects.

>>146
Cândido Chuva Gomes
Established Cândido Chuva Gomes Arquitectos in 1993. Studied architecture at Lisbon Higher Education School of Find Arts and completed his Master's Degree at the University of Évora in Portugal. Has been invited professor at the University of Barcelona from 2002 to 2004 and Catholic University of Viseu from 2007 to 2011.

>>156
José Neves
Was born in 1963, Lisbon. Established his own office after graduating from Technical University of Lisbon School of Architecture in 1986. Has been teaching at the same university since 1990.

>>68
Sid Lee Architecture
Jean Pelland[left] is a cofounder and senior partner of Sid Lee Architecture. Graduated from Université de Montréal in 1995, major in landscape design. Has an experience of 21 years in architecture.
Martin Leblanc[right] is a senior partner of Sid Lee Architecture. Graduated from Université de Montréal in 1994 major in architecture. Started his career in 1991 in the fields of architecture, urban design and interior design.

>>164

Daniel Fügenschuh
Studied at the Technical Faculty of Architecture of Innsbruck, Austria and received diploma "European 4". Concentrates on landscape architecture and urban planning. Won many prizes in architectural competitions. Has taught in the University of Innsbruck and other universities abroad.

>>180

SOMAA + Gabriele Dongus
Is an architecture firm based in Stuttgart and Barcelona founded by Hadi A. Tandawardaja and Tobias Bochmann. Hadi A. Tandawardaja[left] studied at the University of Stuttgart and in Switzerland at the EPFL Lausanne. Has been nominated for the Design Award of the Federal Republic of Germany and was awarded the DDC Award for Good Design. Has been lecturing at the University of Stuttgart since 2007, and at the HFG Schwaebisch Gmuend since 2012 in process design. Expert junior in international architecture competitions. Tobias Bochmann[right] studied at the University of Stuttgart and the EPFL Lausanne in Switzerland. Since 2001, has been responsible for the design of various award-winning buildings and won the BDA "Award for Good Buildings" in 2005. In 2009, began lecturing at the University of Stuttgart. Since 2011, Tobias Bochmann is assistant professor at the Institute for Housing and Design at the University of Stuttgart.
Gabriele Dongus[bottom] studied at the architecture of applied sciences in Stuttgart. Worked for remarkable offices in the south of Germany for administrative and office projects. Later, in her own office, realized a substantial number of projects from detached houses to refurbishments and renewals.

Tegnestuen Vandkunsten
Was established in 1970 by Svend Algren, Jens Thomas Arnfred, Michael Sten Johnsen and Steffen Kragh who are members of the Danish Academic Architects Association.
Covers all the aspects of the architectural profession from landscape and urban planning, housing developments, urban renewal and building renovation to housing and commercial construction projects as well as institutions for artistic or social

>>172

>>20
Silvio Carta

PhD, He is an architect and critic based in Rotterdam. Lives and works in the Netherlands, Spain and Italy, where he regularly writes reviews and critical essays about architecture and landscape for a diverse group of architecture magazines, newspapers and other media. In 2009, he founded The Critical Agency™|Europe.

>>124
Jorge Alberto Mejia Hernández

Had his education as an Architect at the Universidad del Valle and graduated in 1996. Holds a Master in History and Theory of Art and Architecture (2002) as well as a Master degree in Architecture (2008), both from the Universidad Nacional de Colombia. His teaching includes Architectural Theory, History of Architecture and Design Studio at the Universidad Nacional de Colombia since February 2005, where he became Professor Catedratico Asociado in 2007.
Has written many books such as *Enrique Triana: Obras y Proyectos* (Bogotá: Planeta, 2006) and *Coauthor of Vivienda Moderna en Colombia* (Bogotá: Universidad Nacional de Colombia, 2004) and *XX Bienal Colombiana de Arquitectura* (Bogotá: Sociedad Colombiana de Arquitectos, 2006). His research interests include architectural form, modern architecture, contemporary conditions and architectural principles and procedures.

>>106
TETRARC

Was founded in 1988 by Michel Bertreux, Alain Boeffard, Claude Jolly and Jean Pierre Macé (from top left). Has a very broad architectural approach: urban, public spaces, landscapes, new constructions, industrial design, exhibition design and museography. Its four founding members pursue the idea of total art, regarding architecture as poetry. In 2010, Romain Cateloy, Caud Daniel Patrick, Moreuil and Olivier Perocheau (from bottom left) who had been already working in TETRARC were granted partnership.

purposes. Also, they undertake programming, professional supervision, project management and full-service consultancy.

图书在版编目(CIP)数据

旧厂房的空间蜕变 / 韩国C3出版公社编； 时跃等译.
— 大连：大连理工大学出版社，2012.7
（C3建筑立场系列丛书；17）
ISBN 978-7-5611-7093-9

Ⅰ．①旧… Ⅱ．①韩… ②时… Ⅲ．①工业建筑－室内装饰设计－世界－图集 Ⅳ．①TU27-64

中国版本图书馆CIP数据核字（2012）第155118号

出版发行：大连理工大学出版社
　　　　　（地址：大连市软件园路80号　邮编：116023）
印　　刷：精一印刷（深圳）有限公司
幅面尺寸：225mm×300mm
印　　张：12
出版时间：2012年7月第1版
印刷时间：2012年7月第1次印刷
出 版 人：金英伟
统　　筹：房　磊
责任编辑：杨　丹
封面设计：王志峰
责任校对：张媛媛

书　　号：ISBN 978-7-5611-7093-9
定　　价：180.00元

发　行：0411-84708842
传　真：0411-84701466
E-mail: a_detail@dutp.cn
URL: http://www.dutp.cn